Operationalizing Twenty-First Century Safety

A Humancentric Practical Guide

Edited by

Simon Goncharenko

CRC Press
Taylor & Francis Group
Boca Raton London New York

CRC Press is an imprint of the
Taylor & Francis Group, an **informa** business

Designed cover image: Shutterstock

First edition published 2025
by CRC Press
2385 NW Executive Center Drive, Suite 320, Boca Raton FL 33431

and by CRC Press
4 Park Square, Milton Park, Abingdon, Oxon, OX14 4RN

CRC Press is an imprint of Taylor & Francis Group, LLC

ISBN: 978-1-032-95078-5 (hbk)
ISBN: 978-1-032-94980-2 (pbk)
ISBN: 978-1-003-58310-3 (ebk)

DOI: 10.1201/9781003583103

Typeset in Times
by Newgen Publishing UK

Contents

Contributors

Noah Cadieux is Co-Founder and Director of Neurodiversity Programs for Divergent Performance, an organization specializing in advancing neuroinclusion in the workplace. Noah focuses on integrating research findings from the neurodiversity paradigm with the safety sciences and Human and Organizational Performance principles to enhance well-being of neurodivergent workers.

Noah holds a Master of Engineering in Advanced Safety Engineering and Management and a Bachelor of Science in Psychology from the University of Alabama at Birmingham. Noah has served as a consultant, guest lecturer, and speaker on the topic of neurodiversity for the past 4 years, including speaking engagements at the American Society of Safety Professionals Safety 2023 and 2024 national conferences, Board of Certified Safety Practitioner's 2024 Global Learning Summit, and the 2024 US Naval Safety Symposium.

Randy Cadieux is Co-Founder and Director of Operations for Divergent Performance, an organization specializing in advancing neuroinclusion in the workplace. Randy is also Co-Founder and Vice President for Academic Engagement at SPARTAN Training and Performance, where he creates academic collaborations, and develops, delivers, and instructs Human and Organizational Performance courses. Randy is also Adjunct Instructor in the University of Alabama at Birmingham's Master of Engineering in Advanced Safety Engineering and Management (ASEM) program.

Prior to his work in consulting, coaching, and teaching Randy served 20 years in the US Marine Corps, primarily in aviation roles. He was a KC-130 Transport Plane Commander, a UC-12B Transport Plane Commander, and T-34C Instructor Pilot. He is also a graduate of the US Navy's Aviation Safety Officer course and Crew Resource Management Instructor course and served at the director level of safety management in the military.

Randy holds a Master of Science in Information and Telecommunications Systems Management with a Concentration in Project Management from Capitol Technology University, and a Master of Engineering in Advanced Safety Engineering and Management from the University of Alabama at Birmingham. Randy has been consulting with high-risk industries for over 10 years and is the author of *Team Leadership in High-Hazard Environments: Performance, Safety and Risk Management Strategies for Operational Teams*, published by Taylor & Francis.

Jonathan K. Corrado has a professional background primarily in the nuclear industry, where he has expertise in program management, nuclear safety, nuclear criticality safety, regulatory compliance, event investigation and reporting, regulatory commitment management, and oversight of industrial safety, waste management, environmental monitoring, and nuclear materials control and accountability.

Prior to his work in the nuclear industry, Dr. Corrado served in the US Navy, where he was a nuclear surface warfare officer. He continued his naval career in the US Navy

Reserve where he is currently a senior officer and has held a breadth of assignments including command roles on several occasions. Dr. Corrado also worked at a DoD Laboratory performing research and development and in the defense industry managing systems engineering and development of covert communication methods, low size, weight, and power (SWaP) systems, and quick react solutions for the US Intelligence Community.

Dr. Corrado holds a Bachelor of Science in Mechanical Engineering from the Virginia Military Institute, a Master of Engineering Management from the Old Dominion University, a PhD in Systems Engineering from the Colorado State University and is a graduate of the Air War College, Navy Nuclear Power Program, and Naval Postgraduate School. He is also a licensed professional engineer in mechanical engineering in the State of Ohio.

He has research interest in several fields, including nuclear engineering, systems engineering, human performance, safety, theology, and military affairs, and has authored many articles, refereed journal papers, and books in these fields.

I. David Daniels is an occupational health and safety professional, former public safety executive, thought leader, and founder/CEO of *ID2 Solutions, LLC*, a safety-focused solutions company specializing in helping organizations plan and execute safety management systems, including focusing on psychosocial hazard mitigation strategies. He hosts the Psych Health and Safety USA podcast and is the author of a new book, *Psychosocial Hazards Are Real*, which looks at the seminal safety issue of our time, psychological health, and safety. His significant contributions to the field have been recognized with numerous awards and honors.

Dr. Daniels's extensive education and certifications, coupled with his years of experience, make him a trusted expert in the field. He holds a Bachelor of Science in Fire Services Administration, a master's degree in Human Resources Management, and a PhD in Occupational Health and Safety. His certifications as a Safety Director, Violence Prevention Specialist, Emergency Management Specialist, and Job Hazard Analysis Specialist further attest to his expertise. He has also honed his skills at prestigious institutions such as Harvard University and Cornell University.

C. Thomas Goncharenko is a true servant leader. He currently holds the role of the Marketing Manager at TEAM Solutions, a Texas Energy Automation Management company. Prior to this, Thomas worked for Amazon, which recruited him 18 months before he graduated from college. During his time at Amazon, he managed a team of 100 associates at one of the largest fulfillment centers in Houston, TX. Thomas's passion for people, compassionate heart, and keen emotional intelligence enabled him to take his Amazon team to be number 3 in the state, which is no small accomplishment.

Thomas holds a Bachelor of Science in Business Marketing from Liberty University. Thomas has served as a consultant, trusted advisor, and volunteer in many nonprofit contexts.

Simon Goncharenko (PhD, CSP, CSHO) is Head of Expert Services at Veriforce, LLC, and member of the Veriforce Strategic Advisory Board. He is also a Professor at Liberty University. Having been born and raised in Eastern Europe, Simon's 30 years of professional experience has seen him lead teams on three continents and in various industries, including oil and gas, construction, and data centers. His most recent employment prior to Veriforce was with Meta Platforms, Inc., formerly Facebook, where Simon supported the construction of multi-billion-dollar hyperscale AI data centers. Dr. Goncharenko has authored or contributed to over 80 articles, including in the ASSP's *Professional Safety Journal*, and 6 books, including *Save Lives: Pushing Boundaries in Human Factors*. The companion training to the book, entitled *Save Lives Global Human & Organizational Factors©*, has been widely popular with hundreds of management and frontline participants around the globe. The book and training above has also morphed into a global initiative – Save Lives Global – that trains, consults, and advises corporate and government stakeholders in strategies for safer working outcomes (savelivesglobal.com). Simon is an animated and engaging master trainer, charismatic keynote speaker, and podcaster of the *Save Lives Global Podcast*.

Steven Haynes is Director of the Risk Management and Insurance Technology Program and Assistant Professor of Practice in Finance and Managerial Economics at the Naveen Jindal School of Management at the University of Texas at Dallas. Dr. Haynes earned his PhD in Fire and Emergency Management Administration from Oklahoma State University in 2021. He also holds a Master of Public Administration (2016) and a Bachelor of Science in Emergency Management (2011) from the University of North Texas. Prior to joining the University, Dr. Haynes worked for 15 years in the insurance and risk management industry.

His areas of expertise include risk analysis, organizational management and culture, and mixed-methods designs. In addition to his academic roles, Dr. Haynes has contributed to publications such as *Safety Science, Firehouse Magazine*, and the *Journal of Business Continuity and Emergency Planning*, where he focuses on topics related to risk analysis and organizations in crisis. He resides in Allen, Texas, and is a proud US Navy veteran.

James A. Junkin (MS, CSP, MSP, SMS, ASP, CSHO) is chief executive officer of Mariner-Gulf Consulting & Services, LLC and chair of the Veriforce Strategic Advisory Board and past chair of *Professional Safety* journal's editorial review board. James is a member of the Advisory Board for the National Association of Safety Professionals (NASP). He is Columbia Southern University's 2022 Safety Professional of the Year (Runner-Up), a 2023 recipient of the National Association of Environmental Management's (NAEM) 30 over 30 Award for excellence in the practice of occupational safety and health and sustainability, and the American Society of Safety Professionals (ASSP) 2024 Safety Professional of the Year for Training and Communications, and the recipient of the ASSP 2023-2024 Charles V. Culberson award. He is a much sought-after master trainer, keynote speaker, podcaster of The

Risk Matrix, and author of numerous articles concerning occupational safety and health.

Marosh Kulhavy is a seasoned Learning Experience Manager at DataCenterDynamics, where he designs innovative training programs that bridge technical knowledge with practical application. With over 20 years of experience in instructional design and technical training. Marosh is passionate about creating engaging, learner-focused solutions that drive results for individuals and organizations alike. He holds a master's degree in Safety and Quality Production from the Technical University of Košice, Slovakia, and has earned certifications in Lighting Design, Six Sigma Green Belt, and Coaching & Mentoring.

Marosh's journey to becoming a leader in Learning and Development is as unique as it is inspiring. He began his career as a professional dancer, achieving the titles of European and British Salsa Champion. As a certified dance instructor, he developed a deep understanding of discipline, precision, and the power of clear communication – skills he now applies to his corporate work. His ability to translate complex ideas into accessible, actionable insights has made him a trusted expert in the technical and educational fields.

As a published author, Marosh explores cutting-edge topics such as human-centric lighting and the role of human factors in data center operations. His writing reflects his dedication to enhancing productivity, well-being, and innovation through smart, human-focused solutions.

Marosh's unique blend of technical expertise, creative thinking, and a deep commitment to helping others succeed makes him a valuable resource for learners and professionals seeking to excel in today's dynamic world.

Erich Pyles is a passionate and dedicated health and safety leader, adept at enhancing human and organizational performance improvements while collaborating with other leaders to foster sustainable success.

Erich is a seasoned and accomplished environmental health and safety (EHS) and risk management professional, holding a Doctorate in Executive Leadership from the University of Charleston, West Virginia, and a Master of Science in Health and Safety Management from Columbia Southern University in Orange Beach, Alabama. With over 25 years of experience spanning various industries – including the United States Army, heavy industrial construction, oil and gas, utility generation, data centers, semiconductor, and energy sectors – Erich integrates leadership, safety, and technical knowledge through a human-centric approach.

Following his military service, Erich began his career as a Union Boilermaker, where he acquired valuable knowledge and practical skills through a skilled tradesperson (blue-collar) perspective while working in and around power plants. This foundational experience has equipped him with a profound understanding of the human-centric elements of the workplace, enabling him to identify opportunities for improvement among frontline leaders, organizational management, and employees, with a focus on solutions and influencing change.

Erich's demonstrated capability to lead teams in hazardous and non-hazardous environments, combined with his expertise in business strategy, EHS, leadership coaching, employee engagement, and organizational development, has allowed him to support others to excel in today's ever-changing work environments.

Todd Shilling is a pracademic – a practitioner and an academic. As an academic, he is a Doctoral Chair and External Reviewer for Capitol Technology University. As a practitioner, Dr. Shilling currently works in the offshore oil and gas industry as a Project HSE lead. His knowledge of safety and health requirements comes from 20 plus years of experience with an extensive background of working in multiple international and domestic locations. Todd is also a 30+ year Nationally Registered Paramedic.

Introduction

Most people skip this part. Why waste precious time on introduction when you have rich content to get to? I get it, we are all extremely busy and each of us has a variety of different tasks vying for our time and attention. So, if you are here and you are actually pausing long enough to read this, I will respect your time and effort and make this brief.

First, thank you for giving us your attention. We don't take it for granted and want you to know that we appreciate you.

Second, if the title or the description of this book caught your initial attention, let me tell you why it would be worth your while to read this book in its entirety. When I got the idea for my own book entitled *Save Lives: Pushing Boundaries in Human Factors,* I did something that almost by accident led me to compiling the collection you have in your hands today.

What did I do?
I broached the idea for an easy-to-read practical manual about the human and organizational factors approach to safety, meant to take Conklin's, Dekker's, Reason's, Hollnagel's, and others, research to the next more practical level to a number of my colleagues, friends, and LinkedIn connections. These individuals whom I approached were all experts in their field, highly respected, and extremely knowledgeable. So, I valued their feedback. They were all very enthusiastic about my idea. I was happy to see this but what happened next surprised me. They were also all offering to contribute a chapter to my book. I had not expected that.

I reflected on their feedback and their offers and came up with the perfect game plan. Step 1 was to go ahead and commit to paper the thoughts and ideas that were circulating in my own mind into a one-author book. That title, referenced above, was published in the early part of 2024 and is now available via Amazon. Step 2 was to humbly accept their offers and become an editor of an amalgamation of their brain power, experience, and industry know-how, which is what you have in your hands today. *Operationalizing Twenty-First Century Safety: A Humancentric Practical Guide* is the product of our collective work.

Third, this brings me to the point of what makes the book you're holding now unique. The most obvious part – it probably has the longest title of any book you

DOI: 10.1201/9781003583103-1

read before. But what really stands out about it is the people who put it together. Who are we?

- We are practitioners and academics – you know what they say about those who can't do? They teach. Well, we do and teach about what we do.
- We come from a variety of different industries, including nuclear energy, construction, data centers, oil and gas, education, etc.
- We are experts in our respective fields, with many of us being previously published authors, sought-after speakers, master trainers, global consultants, and invited to sit on organizational and industry standard boards, like the American National Standards Institute and others.
- We are multigenerational, so can relate to your values, interests, pain points, and concerns.
- We are unique in our multinational and multi-lingual backgrounds – hailing from multiple continents.
- We have over 300 years of combined industry experience.
- We are veterans, proudly serving our respective homelands.
- We are neurodivergent – finding our ways and leveraging our understanding to help you succeed.

So, I think it is safe to say that in reading this book to the end you will learn, grow, get ideas, have your creativity stirred, find answers to the questions you didn't know you had, and become a better human being as a result.

When you do, reach out to us and connect with us on LinkedIn. Tell us your honest opinion because we actually do care.

Enjoy the journey ahead.

Dr. Simon Goncharenko
Editor

1 Construction Site to Cubicle

Human Performance Tools for All Walks of Life, as Informed by the US Nuclear Navy's Success

Jonathan K. Corrado

Responsibility is a unique concept. ... You may share it with others, but your portion is not diminished. You may delegate it, but it is still with you. ... If responsibility is rightfully yours, no evasion, or ignorance or passing the blame can shift the burden to someone else. Unless you can point your finger at the man who is responsible when something goes wrong, then you have never had anyone really responsible.
— **Admiral Hyman G. Rickover,** *"Father of the U.S. Nuclear Navy"*

FULL STEAM AHEAD!

The United States has used nuclear power, the world's most misunderstood source of energy, to fuel its most capable warships for over half a century. Having built over 200 nuclear power plants, the US Navy presently operates approximately 100 operational reactors, which is roughly equivalent to the number of operational civilian reactors in the United States.

With over 20 years of service as a commissioned Navy officer, I have lived and breathed the principles about to be shared with you in this chapter. So, the information below is not just theoretical research but a set of well-practiced principles that brought about operational and safety results desired and looked up to by every single industry and organization around the world. So, what does the Navy do that sets it so far apart and can it be emulated?

A naval reactor, in contrast to most commercial reactors, does not commence operations and maintain full power for extended periods of time until it is necessary to refuel. Rather, these reactors are subjected to a wide range of conditions, transported to both great depths and the far reaches of the Earth, and frequently subjected to atypical conditions or cycled for the purpose of operator training. With that said, it is not uncommon to shut down and start up a naval reactor two or three times per day, and between periods of operation, place the equipment in a variety of unconventional operating configurations to train and qualify those who will use it in military scenarios.

Despite the arduous conditions in which these reactors operate, the US Navy has succeeded in avoiding a radiological accident over the long life of its nuclear power program. Although there have been substantial maritime incidents, such as the sinking of the submarines *USS Scorpion* and *USS Thresher* (which have been on the ocean floor for nearly 60 years), reactor plant casualties were not the cause of these incidents, and despite their outcomes, none resulted in a meaningful radiological hazard to people or the environment.

How is it, then, that the US Navy has preserved such an immaculate safety record, despite the austere operational conditions and large number of plants operated? Is it the operators' years of training and certification? Is it substantial regulatory oversight, inspection, and accreditation?

Not really.

The Naval Nuclear Power Program generally recruits high school students and certifies them to operate a nuclear power plant in less than two years. From "irresponsible" teenager to a nuclear power plant operator in two years? Can you imagine anything like that on the civilian side?

Me neither.

Maybe the secret is in their supervision, you may think. These individuals are recent college graduates who, for the most part, possess significantly less experience than the individuals they supervise. On the surface (no pun intended), this may appear to be a recipe for disaster, so how, then, can the program's effectiveness be explained, and how can others leverage this success to benefit their organizations? Surely, there is more to it than just the difference between the surface and the water…

Believe me, there is.

Everything—the grueling naval nuclear power training program, nuclear service in the fleet, the effective operation of the nuclear reactors, and the crew's safety—depends on everyone adhering to seven very important watchstanding principles that are ingrained in every nuclear operator's core. Adherence to these principles has resulted in the safe operation of naval nuclear propulsion plants for more than 70 years, without even a single reactor accident or inadvertent release of radioactivity that was dangerous to human or marine life.

Naval reactors have an outstanding record:

- Over **166 million miles** have been safely steamed on nuclear power!
- Over **7,100 reactor years** of safe operation have been amassed!

Every industry would love to emulate those results. Every organization is eager to make similar advancements in their efforts to decrease human error and eliminate fatalities, injuries, and incidents. But is every organization willing to go the distance that the Navy has gone in establishing the systems that produced its impressive results? This is the million-dollar question.

And before you jump to answering it, read on, so you can be clear on what the total commitment truly entails.

The watchstanding principles are:

1. Integrity,
2. Ownership,
3. Procedural compliance,
4. Formality,
5. A questioning attitude,
6. Forceful "watchteam" backup, and
7. Level of knowledge.

These seven principles steer all aspects of naval nuclear power program operation and are the criteria by which all operators are judged.

Over time, the success of the Navy's nuclear power program has gained traction. Many organizations have adopted these principles, with a significant return on investment. Readers may recognize several of these principles and may currently be using them, but hopefully, this chapter will share some new insights that can be incorporated into other safety cultures.

Additionally, each principle is not independent. Like the foundation and frame of a home, they all provide mutual support and, as readers will observe, management is essential to the success of the principles.

INTEGRITY

Integrity must be at the foundation of everything—this is why it is listed first. It is more than simply doing the right thing when no one is watching. It means speaking up when a mistake is made and creating a culture where an organization feels safe to do so, as well. That gets into Psychological Safety.

Within the Navy vernacular, "integrity" is synonymous with "honor"—one of the Navy's core values. The colloquial definition that is often used in the service is "doing the right thing even when no one is looking." In Navy life, sailors are taught that, no matter what, whether one is being supervised or whether somebody would ever come to know the truth, one is expected to do *the right thing*, in all facets of life.

In all facets of life is noteworthy too because it goes beyond the immediate battle station to include personal choices and decisions at home, with relationships, etc. Nothing is off limits. Civilian world could benefit from diving deeper and picking apart this principle a little more, in order to find things worth emulating.

Before addressing implementation of integrity, a discussion of what "right" means is necessary. Oftentimes, the distinction between "right" and "wrong" is straightforward, and this standard is universally held. However, occasionally, the definition of "right" fluctuates. Following some violations of integrity, the Navy decided to redefine the term and the manner in which it was implemented.

Navy leadership determined that the definitions of "right" and "wrong" were contingent upon a combination of the individual's moral standards, cultural upbringing, and the situation in which they found themselves. They also determined that the implementation of a zero-defect organization, and more specifically, the way leadership enforced that ideology, led to what is known as the "normalization of deviance." The sociologist Diane Vaughan coined this term to denote the process by which deviance from correct or appropriate behavior becomes acceptable in a culture. This phenomenon may manifest in organizations, groups, or individuals, and it may result in diminished safety or performance standards. Vaughan used the term when analyzing the *Challenger* space shuttle disaster, which was attributed to overlooking or rationalizing flaws in the shuttle design.

In organizational life, the normalization of deviance means what is considered *acceptable*. This can be as straightforward as neglecting to buckle one's seatbelt before driving a car or as hazardous as neglecting to conduct equipment safety checks prior to reactor start-up. The reader may now wonder: Who condones this level of deficiency in standards? For leadership, management extends beyond merely the enforcement of standards. One must ask: What is an employee's rationale for deviating from the prescribed rules? Does the employee not fully understand them? Are the rules inadequate for the given task? Are there external pressures or expectations that force the employee to deviate? Is there a barrier, such as an unrealistic expectation for accomplishment, that causes the employee to feel justified in circumventing the rules?

The Navy discovered that a combination of these factors resulted in what leadership regarded as their most severe integrity violations. To reorient themselves, they issued a revised definition of integrity: "Absolute honesty, trustworthiness, and reliability, in … training, qualification, operations, and maintenance. Demonstrating moral courage to accept responsibility for one's actions." This definition was built on what was advantageous for the organization rather than the issue of morality. It established a shared sense of accountability: The individual was accountable for fulfilling their obligations, while management was responsible for ensuring that the expectations were clear and reasonable. To implement this concept in other organizations, it is necessary to establish a culture of *right* that is centered on the organization's interests, rather than those of any individual.

Let us explore an example. A piece of equipment suffers a bearing failure, halting the entire operation. After the failed part is replaced, it becomes apparent that the bearing had not been lubricated. How should management respond? One way would be to fire the employee responsible for failing to perform the maintenance. However, one should ask: Would that solve the problem? Upon assessing the situation, leadership learns that the employee assigned to the task, which normally requires two days, was only given one day to do it. Additionally, they learn that the site did not have the required grease for the maintenance activity. Finally, after reviewing the procedure, leaders realize that the piece of equipment was modified, affecting the bearing that had the failure. However, the maintenance list and drawing were not updated. Given these findings, does firing an employee fix the situation or at least identify a larger problem? Clearly, firing the employee does not solve the root cause of the problem, which would likely happen again.

Institutional integrity drives an understanding between management and employees that each individual is responsible for identifying and correcting deficiencies in the process, whether they are procedural, material, or personnel related. To be effective, management must ensure that organizational expectations are reasonable and well-communicated, and that any issues that prevent employees from meeting these expectations are taken seriously and resolved. And this must be repeated and reviewed on an ongoing basis.

Institutional integrity prioritizes the organization's and its constituents' welfare over personal advantages (including safeguarding allies). This approach to integrity is significantly superior to imposing the responsibility to behave morally on the individual, irrespective of the negative circumstances that the organization may impose upon them. Additionally, it promotes the next principle.

OWNERSHIP

Admiral Hyman G. Rickover, known as the "father of the U.S. Nuclear Navy," once stated,

> Responsibility is a unique concept. ... [Y]ou may share it with others, but your portion is not diminished. You may delegate it, but it is still with you. ... [I]f responsibility is rightfully yours, no evasion, or ignorance, or passing the blame can shift the burden to someone else.

While the definition of "ownership" is somewhat subjective—similar to that of "integrity"—some individuals arrive at work each day with the mindset that they are essential to the success of this team, and others are content with performing solely the tasks outlined in their job description, without going above or beyond those responsibilities. Although neither employee is inherently incorrect in their approach to work, one employee is clearly preferable to the other.

Ownership extends beyond productivity and influences each principle; all the elements fall into place once ownership is established. To impress ownership on an organization, leaders should establish a culture in which all employees are committed to the organization's success and errors are readily rectified. Build an environment in which all employees feel essential to the organization's success and assert, "If any of us are wrong, it is incumbent on all of us to point it out." This fosters a sense of ownership.

For example, a team member who possesses the watchstanding principle of ownership walks thorough the workplace and notes areas that require cleaning, that have unfavorable working conditions, or where training is urgently needed. On a daily basis, they ask, "How can I aid in improving this organization?" Nothing is someone else's responsibility. If I can do something about it, I will.

PROCEDURAL COMPLIANCE AND FORMALITY

Although procedural compliance and formality are discrete watchstanding principles, they share a theme and desired outcome. In practice, procedural compliance and formality can be oppressive and create a slower-than-desired pace of task or operation completion. However, this is precisely the purpose of these principles, which aim to reduce human error.

First, procedural compliance concerns ensuring a thorough understanding of processes so they can be performed with consistency. In an operational context, this does not necessarily need to manifest as a robust set of procedures with step-by-step directions. The key is to provide the context and the *why* behind what employees are asked to do, so that they have this understanding. This ensures consistency of instruction and understanding among all employees. It also empowers employees to make choices that are in line with the intent of the process, even if there is no formal documentation.

Like procedural compliance, formality may come across rigid and stuffy, but it simply concerns the *consistency* of practice. It ensures the preservation of the culture created via adherence to the other watchstanding principles. It provides employees with an understanding of the right way to do things and why they should do them in that way.

People enjoy the flexibility and freedom to be themselves, which is also certainly encouraged. This principle can be applied simply by ensuring that all employees meet a certain standard regarding the other principles. As a leader, if something seems amiss, the employee should be informed, and a course should be charted to identify how to address that gap and grow, while providing the context and the *why* behind these actions. Ultimately, human error is mitigated by procedural compliance and formality. Those who do not know what they do not know are the most hazardous in an industrial setting; they are unaware of their own vulnerabilities.

Formality and procedural compliance ensure that all parties are in agreement. It is possible to prevent significant issues caused by a gap in an employee's level of knowledge by using common terms in communication, giving plain and concise instructions, and adhering to written procedures. Conversely, people are not impervious to error. An employee can question and provide forceful support if they possess the necessary knowledge and a questioning attitude. In turn, management is accountable for assessing the concern and offering a rationale for its accuracy or inaccuracy, as well as instruction on how to proceed (ownership).

Sometimes, a workplace change is made, equipment upgrades are installed, or procedures are written that simply do not work. If procedural compliance is a core value, and the employees have ownership, they will bring such issues to management's attention. Nevertheless, if an employee develops a workaround but fails to send it up the chain of command to ensure that the procedure is corrected, or—even worse—if the failure of a component is the cause of a procedure's inadequacy, then nothing is resolved. At some point, the employee will share the shortcut with another employee, who may develop their own shortcut and may or may not comprehend the rationale behind the initial shortcut or the definition of "right." In the most favorable scenario, a company simply loses money due to inefficiencies. In the most serious scenario, an incident occurs that results in injury, the loss of life, equipment damage, or catastrophic loss.

The solution to the problem of procedural compliance above can be resolved through formality. The expectation is that if the plant is operated in a formal way that complies with written procedures, any deficiencies in those procedures or equipment will be identified and forcibly communicated to management to develop a solution. Ultimately, all parties benefit from the resolution of issues and the enhancement of procedures.

QUESTIONING ATTITUDE

A questioning attitude means not being complacent about why things are the way they are. It means not blindly accepting a process because "that's the way we've always done it." Having a questioning attitude creates a sense of mastery and assists in employee's learning and development. A culture where employees feel encouraged to speak up and ask questions allows them to develop a better understanding of the subject matter (i.e., mastery). Again, certain level of Psychological Safety must exist in this culture. This also improves knowledge and processes by purging inefficiencies. This attitude resembles the notorious "hunch" that is common in popular culture. The Navy refers to this as a "questioning attitude," and it extends beyond the simple inquiry, "What will occur next when I execute this action?" Employees must consider this notion each time they press a button, operate a valve, or adjust a setting; however, there is also more to it.

In contrast to merely anticipating the plant's response to an action, the questioning attitude taps into an employee's "spidey sense" when something is about to occur and simply does not feel right, even if it is not specifically within that employee's area of concern. This is achieved by linking the level of knowledge and forceful "watchteam" backup (the next principles to be discussed).

Employees can effortlessly implement a questioning attitude; however, managers must exert considerable effort to cultivate it. A significant amount of perseverance is necessary. Supervisors may be compelled to address hundreds of inquiries that ultimately do not address actual issues. It may get old and annoying rather quickly. But interrupting the development of this attitude is very short-sighted. Persevering with patience in the development of questioning attitude fosters employee development, increases their sense of ownership, and—when a serious issue is eventually identified—prevents a costly mistake. So, it very literally saves your behind!

FORCEFUL "WATCHTEAM" BACKUP

As previously mentioned, none of these principles is autonomous, and management is essential to their success. An organization's leadership establishes its tone. With that said, the most significant obstacle associated with this principle is the tone through the use of the word "forceful."

In the Navy, "standing watch" basically means being on shift—being on the watchteam that is responsible for the entire ship's function during the shift. The concept of watchteam backup concerns empowering all members of the team to speak up when they see something that does not look right. Providing good watchteam backup can be difficult, and many are not used to this practice. If done incorrectly, people can receive it as a criticism or as an indication that a person does not know how to perform a task correctly. No one wants to look foolish or uninformed.

Again, however, the goal is *empowering* employees and ensuring that they can make the best possible decisions given the information provided. The concept that "two heads are better than one" applies here. If someone tries to move an idea forward, but another employee (even one with less experience) has more or different context that could change the approach, they should feel empowered to speak up. To avoid hurt feelings, this should be made part of the standardized plan for the workforce. Such questioning should be expected, rewarded, and praised. Watchteam backup also teaches employees to respect each other's strengths and rely on one other.

This principle has a dual application. First, if an individual knows or perceives that something is amiss, they are obligated to speak up and have the authority to press the matter. Second, superiors are accountable for accepting feedback with an open mind and ensuring that all team members agree that an issue has been adequately addressed for the team's benefit before proceeding.

To derive the full benefits of this principle, it is imperative that the entire organization abandon their egos. Easier said than done, right? Applying this principle begins with soliciting feedback from employees, and regardless of the nature of that feedback, all concerns must be addressed before proceeding. Additionally, it is highly likely that, if one employee raises an issue, another has had the same issue before but was hesitant to raise it.

Implementing forceful backup is challenging. For instance, a junior employee may be apprehensive about speaking up, in case they are proven incorrect. Likewise, senior personnel may become overconfident in their own knowledge and may be offended if they are questioned. Such scenarios cripple the application of forceful backup, as Rickover suggests: "Free discussion requires an atmosphere unembarrassed by any suggestion of authority or even respect."

Open dialogue based on technical correctness is permitted; however, insubordination is not excused. Aligning institutional integrity, ownership, and the level of knowledge with the principle of forceful backup and maintaining the culture's emphasis that "It isn't about *who* is right, it is about *what* is right" makes it possible to capitalize on open, candid dialogue. In organizations that embrace such dialogue, it is astounding the information uncovered that can be acted upon to make improvements in not only safety and human performance, but also efficiencies in productivity and cost savings. Forceful reinforcement reveals all parties' true levels of knowledge, which is advantageous for all, reduces errors in the workplace, and ultimately promotes a culture of active safety consciousness.

LEVEL OF KNOWLEDGE

The team's welfare and safety are the primary objectives of all actions. In some organizations, leaders expect employees to execute their directives without hesitation. This is how the military operates. In contrast, the nuclear Navy operates differently. Admiral Rickover notoriously harbored a deep-seated aversion to the rigidity of military organization. He believed that blindly adhering to another's decisions, a practice that was exclusively based on longevity, was perilous. In Rickover's view, the value of one's qualifications and knowledge was more significant than their level of service or rank. To this end, Rickover was preoccupied with the concept of *level of knowledge*, and his program was predicated on maintaining the highest standards for operators. The "maintenance" of individuals is contingent upon their level of knowledge, and to function as expected, individuals require preventive and occasionally corrective maintenance, similar to machines.

The level of knowledge is the bedrock of the organization. If the workforce does not know what to do or why, they cannot be effective. In fact, a lack of knowledge can cause mistakes and lead to serious consequences. This principle is also the foundation for other principles. How can an employee have a questioning attitude if they do not understand the environment? Can they provide backup if they do not know the right answer? Management must cultivate an effective workplace environment to ensure that employees possess a base level of knowledge to perform their job correctly and to the standards that their clients expect.

Effective training programs should provide the fundamental skills that employees need prior to their involvement with equipment. Additionally, such programs should offer beneficial remediation for noncompliance with these standards—not in terms of discipline, but rather in terms of overall success. A meaningful plan should be developed to address the fundamental cause of issues that reveal a lack of knowledge in a particular area. The most significant benefit is that the combination of a culture of ownership and a high level of knowledge will motivate employees to develop their intelligence beyond the training they receive. This is instrumental! Developing a culture of learning within organizations enables teams to leverage the best of each person's knowledge and abilities.

Not surprisingly, in some workplaces, low morale is accompanied by a lack of knowledge. Few individuals are willing to expose themselves as incapable in front of their peers, and no one wants to be considered incompetent. This outcome indicates a lack of motivation and, even more detrimentally, may result in a normalized deviation in the integrity of an organization's qualification program. The knowledge that the organization deems necessary for the workforce must be imparted without making any assumptions regarding their prior knowledge. It is imperative that workplace cultures foster the exchange of knowledge and vulnerabilities. Training and professional development should be integrated into all employees' daily routines, regardless of their position. This will yield substantial advantages for an organization.

HOPE IS NOT A STRATEGY

As a leader, do you hope that your employees follow accurate, current procedures, or do you *know* that they have the training and tools they need to do so?

Do you hope the workplace is properly maintained, or do you know, based on data and expected results, that proper maintenance is being conducted?

If a leader cannot provide a confident response to these challenging questions, they must assume responsibility and evaluate their organization's deficiencies. It is necessary to consistently assess whether an organization fulfills its objectives safely and optimally. The watchstanding principles are effective, but only when they are implemented properly. However, everything in nature inevitably finds and takes the path of least resistance, and organizations are no different. Management must consistently compare their organization to the desired outcome and make adjustments to reduce the variance.

The two factors that most contribute to the ineffectiveness of these principles are personnel attrition and complacency with an organization's success. In this regard, the nuclear Navy's challenges are not significantly different from those that any other organization encounters: the influx of new personnel, the potential for senior personnel to become complacent, and the retirement of experienced employees. Ultimately, operating a multimillion-dollar nuclear plant silently in the deep ocean and providing a product or service to a customer have the same objective: becoming the most effective organization possible. The Navy has benefited greatly from these principles, and they can be successfully implemented in other organizations with the right approach.

BIBLIOGRAPHY

Grady, John. "Newly Declassified Report Shows How Rickover Worked to Explain Radiation Risk from USS Thresher Loss." *U.S. Naval Institute News* (September 2022). https://news.usni.org/2022/09/08/newly-declassified-report-shows-how-rickover-worked-to-downplay-radiation-risk-from-uss-thresher-loss

Kingsbury, Paul. "When Cheating Becomes Normal." *Proceedings* 141/9/1351 (September 2015). www.usni.org/magazines/proceedings/2015/september/when-cheating-becomes-normal

Meier, Derek. "Principles-Based Operations: A Military-Proven Method Part I." *Power Magazine*, January 4, 2021. www.powermag.com/principles-based-operations-a-military-proven-method/

Meier, Derek. "Principles-Based Operations: A Military-Proven Method Part II." *Power Magazine*, January 4, 2021. www.powermag.com/principles-based-operations-a-military-proven-method-part-ii/

Oliver, Dave. *Against the Tide: Rickover's Leadership Principles and the Rise of the Nuclear Navy*. Annapolis, MD: Naval Institute Press, 2014.

Price, Mary R. and Williams, Teresa C. "When Doing Wrong Feels So Right: Normalization of Deviance." *J Patient Saf.* 14, no. 1 (March 2018). https://doi.org/10.1097/pts.0000000000000157

Rockwell, Theodore. *The Rickover Effect: How One Man Made a Difference*. Annapolis, MD: Naval Institute Press, 1992.

U.S. Department of Energy. *The United States Naval Nuclear Propulsion Program* (2020). www.energy.gov/sites/default/files/2021-07/2020%20United%20States%20Naval%20Nuclear%20Propulsion%20Program%20v3.pdf

Willink, Jocko and Lief Babin. *Extreme Ownership: How U.S. Navy SEALs Lead and Win*. New York: St. Martin's Press, 2015.

Wilson, Paul F., Larry D. Dell, and Gaylord F. Anderson. *Root Cause Analysis: A Tool for Total Quality Management*. Milwaukee: ASQ Quality Press, 1993.

Winnefeld, James A. "Safety Is a Matter of Principles." *Proceedings* 144/2/1380 (February 2018). www.usni.org/magazines/proceedings/2018/february/safety-matter-principles

2 A Worker's Perspective on Human Factors

Todd Shilling

INTRODUCTION

You might have heard an employee on your site say something like: "Those guys in the office have no idea what's going on out here." Or this: "Those rules are made to cover management's butt; it's not about keeping me safe."

The basis of this chapter is to address the seeming disconnect between "the rules" (aka HSE policy/procedure) and the employees that are expected to embrace and utilize the rules in their everyday functions. While this disconnect may not be present in some organizations, it is a serious problem for others, many of whose personnel believe that a good number of the rules are silly, unreasonable, or simply not feasible. Discourse is normal and healthy, but for those that do not believe there is a disconnect or a problem with long-held views on worker safety, it is time to step back, take a look at the everyday operations of your organization, and go have those conversations that will allow the employees to voice their concerns. The whole reason the safety profession exists is built on the premise of keeping the employee safe...... . that simple!

I have been fortunate in my life and career to have had a varying domestic and international exposure to some atypical combinations of work and education that is not normally seen. I am a tenured boots-on-the ground HSE professional with a strong academic background. I have also worked as an offshore facility operations supervisor before returning to the HSE field, where I currently work in a rotational offshore schedule with subsea project teams. So, my words ring true with the guys and gals out there working, as I have been there and occasionally had the same or similar opinions.

Why does this matter? What's the big deal? We all know that safety rules are "written in blood" and they are created in the best interest of the employee to keep them safe, right? Unfortunately, this may not always be the case. While the above is certainly true in many cases, i.e., it is easy to see how wearing fall protection when you're six feet or higher off the deck is critical, some rules and requirements are not so clear or as simple as it may appear at the surface.

What if you're working seven feet off the deck, and there are no appropriate tie-off points, and to try and create a "safe" task, you would create considerably more risk? Now, because the employee chose to do the task without creating more risk (i.e., to "work safe" and tie-off) they are in violation of safety policy which can then lead to a

DOI: 10.1201/9781003583103-3

number of undesirable effects. These are the types of issues that the everyday working person encounters all too frequently. And therein lies the answer to the question: why does this matter?

When discussing the human factors of safety, you will see many opinions, both academic and personal, some derived from pure science, and others from tenure on a job being exposed to many variables. As we shift our unplanned event investigation mindsets from blaming the person to understanding the breakdown within a system that caused an event, these scenarios become even more relevant to the discussion.

Why did the person fall? They were not wearing fall protection. Easy! Right? Now dig deeper, why were they not wearing the fall protection? See above.

This is now starting to scratch the surface and bring the issues of human factors and blanket safety policy matters into relevance. While typically well intended, you may now start to understand how these blanket policies sometimes miss their mark.

Hindsight is 20/20, as the adage goes…Unfortunately, as true as this statement is, we as safety professionals and operations supervisors tend to default back to this mindset. The goal should be to have a robust system that allows us to say that we knew there might be issues and therefore we did such and such.

What is one thing humans are really good at? We make mistakes! Luckily, humans will typically learn from their mistakes, even at the youngest age. I'm going to touch that even if someone said don't. Oh #@*% that was hot, and I just burned my hand. Sound familiar? Is that an incident from someone's childhood, or did that just happen on your jobsite?

What exactly are the human factors in the workplace? According to Vogt, J. et al. (2010), human factors in safety encompass "all those factors that influence people and their behavior in safety-critical situations". While there are many variables on this definition, this one is directed specifically at safety management. When breaking down this one statement, how many factors can and/or will influence employee behaviors throughout the duration of a workday? We can guess the right answer, try to come up with a scientifically calculated hypothesis, or some other form of formula or theory to figure out the number of factors that influence behavior, but the reality is that we don't know, and most likely never will. While we cannot read minds, we do have the capacity to anticipate some of the factors that may influence the employee's behavior, and this is where we typically concentrate our efforts, to one extent or the other.

But what about those influential factors that we cannot anticipate or predict? Do we just move on and hope for the best, create a blanket policy to "CYA" the company? The employee's child was sick and kept them up all night, vomited on their clothing this morning causing them to be late because they had to change and find a sitter, and they couldn't call off because they can't afford to take the loss of pay… sound familiar? Well, company policy says that if an employee is stressed or having family issues, or somehow otherwise distracted, they should take the day off. So, suddenly this person isn't paying attention due to being distracted and twists an ankle creating a medical treatment case or worse. We say the employee was distracted, but we will forgive them and not write them up and require remedial training on slip & trip hazards and wearing proper footwear when they return to work.

While I have no intention of delving into appropriate disciplinary actions for safety infractions, I will be touching on the question of why we look at administrative discipline as a way of correcting/preventing unsafe behaviors and actions. I will also be discussing the opposite, safety awards, "zero" policies, when is enough days without an incident enough, safety culture, and other various recognition schemes that companies have traditionally used to try and boost "safety performance". There will be no condemning of any system or policy but rather discourse and hopefully thought-provoking commentary as it relates to the subjects.

SAFETY INCENTIVE PROGRAMS

Congratulations! Your facility has worked for a year without a recordable incident and for this wonderful accomplishment we are going to have a pizza party and give every employee a company logoed T-shirt! But, in the same breath, we have been seeing a drop in behavior-based safety reporting and this concerns us because we don't want the facility to stop thinking about safety and start working unsafe. While this may not be the exact scenario you have encountered, I'm sure many of you have either been a witness to or have been the company safety representative saying this. For those of you that partake in these types of programs and scenarios, do you truly believe in them or is it more of something that has been forced on you by upper management? If you believe in it and it is working for you, then I am not going to try and change your mind or belief, that's the last thing I want to do, but at least hear me out.

These types of incentive programs are akin to Pavlov's dogs (Mcleod, S., 2024). For those not familiar, Ivan Pavlov was a Russian experimental neurologist known for his discovery of classical conditioning through his experiments with dogs. Pavlov would start out by playing sound to the dogs with no other action. Then he would make the same sound and present food. He then started making the sound without food and the dogs would automatically salivate at the sound. The dogs were conditioned by a specific stimulus.

I'm not saying that your employees are dogs or that we treat them like dogs, but I am simply trying to bring context to how some of traditional safety rewards may simply be conditioning our workforce and not actually improving their safety. We tell them to work safely, and once they hit some arbitrary milestone, we provide them with a reward. Now 99% of the workforce does the same thing for the next specific timeframe, but one person has a bad day that results in an accident, and we take away the reward for everyone. See where I am going with this?

Are we unknowingly creating a "don't know, don't tell culture"? If an employee sprains an ankle but doesn't tell their supervisor, the supervisor has plausible deniability, and the employee just hopes no one will notice they are walking around with a limp for the next day or two. Although the supervisor can clearly see the employee limping, an assumption can be made that it happened during their off-hours. So, the supervisor may ask a benign question on whether the employee is able to work, and of course the employee will say yes. Eventually the safety rep or higher management comes around and notices the limp and prods them until they finally admit to hurting themselves at work yesterday. When both the employee and supervisor are asked why the incident wasn't reported, the supervisor can deny knowing it and the employee states, "I didn't want to ruin our safety bonus!" The employee is given some form of administrative discipline, required to be "re-trained" on various subjects, such as proper foot placement, slips & falls, incident reporting, and whatever else they can drum up to keep them occupied while they are technically still working but assigned administrative tasks for the next day or two until the ankle gets better.

I personally have seen the above scenario played out in various forms multiple times. The employee was never required to seek a medical professional's evaluation/ diagnosis or required to complete a return-to-work medical evaluation. The incident was never reported, and there was never a lost time, work re-assignment, or other cat-egorization officially assigned to the incident, since it was never assigned an actual report. And why did this happen this way? In the name of not ruining a safety record. The random number has been assigned to demonstrate that a facility can work safely by not having an "incident."

This scenario also brings into question whether safety records are a viable source of understanding of how "safely an organization can work." Another aspect highlighted by the scenario above is the common assumption that not reporting an incident would somehow be for the greater good of the rest of the employees. I am not saying that we shouldn't be reporting incidents, nor am I saying safety reward programs are totally bad. But I believe we need to take a step back as an organization and take a deep look at what the objective is and if the current practice is helping to achieve that goal or hurting organization by creating the drama demonstrated above and essentially achieving the opposite of its intended target.

ZERO INCIDENTS

How does this mantra and mindset help if you are not actually doing everything possible to stop incidents? And is that goal even possible? I know I know, if our employees are working safely, following the rules and doing what's right, there is no reason for an incident to occur, blah, blah, blah, blah!

In a perfect world, sure. In reality, one thing is certain, people will make mistakes for a multitude of conscious and unconscious reasons. Why do I say this? Go back and read the first couple of paragraphs! No matter how "good" your safety manage-ment system (SMS) is, people will make mistakes, it's a documented and established fact. Rather than talking ourselves into believing that a company, facility, crew, what-ever is infallible if they follow the incredible and amazing International Standards Organization, Occupational Safety and Health Administration, or whoever approved SMS, we should be asking the question, where can people make mistakes and how can we recover from those mistakes? I have seen large multinational companies that had incredibly detailed, thorough, intricate, well thought-out SMS that took a consid-erable amount of time and resource commitment to create, yet whose jobsites made one quickly realize that all their fancy policies are just a bunch of worthless words on paper to demonstrate that the company cares.

Empty words on paper is what a "strong" SMS is if there is no follow through at the worksite and the company believes that the "paper' (aka SMS) will somehow shield them from having incidents. Now I know some of you are reading these words right now and probably thinking something to the effect of "I worked at so and so and we went 3 years without a recordable incident or an LTI, or my facility went a year without a single incident." Kudos, if this is the case! But, was that really the fact? Anytime I see someone publish, on a popular social media site, that their site worked X amount of time without an incident, I immediately become suspicious. Drawing from the decades of my own observations and experience, I highly doubt you had zero incidents. What was much more likely to have happened in your workplace would be incidents that went unreported, creatively reclassified, ignored, or simply "swept under the rug."

Some may wonder why I take such a cynical view of this. First, read what I wrote above and really what the other authors in this book are saying about human fallibility (we are all prone to make errors all the time and none of us is infallible or free from errors). Second, I know firsthand the harsh realities of working people, especially in heavy industrial environments, and even more so in an international work site. The same could be said for the domestic medical care industry, so it is not isolated to a specific environment or environments. Again, while some industries may have more clear evidence of these realities, the undeniable fact is that all industries which employ people share in the experience. As the old saying goes, "Shit Happens," regardless of how hard you try and work to do things right and follow the rules.

Pardon my digression. I think I made it fairly clear by now that zero incident mentality doesn't work. Things will happen, mistakes made, equipment breaks or malfunctions, or someone simply has a bad day. It's really that simple, things will happen that are essentially out of your direct control sometimes and we must understand and embrace this mentality, nobody is perfect, no matter what slogans or programs you envision. Remember that we are discussing human factors, and as long as there are humans in the equation, mistakes will be made.

This is not to say that I am proposing eliminating the workforce and replacing it with automation either. Taking people out of the equation of the workplace and bringing in automated solutions instead may help in some circumstances, but it will not completely erase mistakes, especially if you have human machine interaction. While I will not delve into this specific topic, I will say that introducing machinery/automation and humans into the same equation will automatically double your risk profile.

SAFETY CULTURE

Let's segway for a second and discuss the infamous "safety culture." What is safety culture? It is this grand and mysterious thing that we somehow seem to conjure up to try and describe how safely or unsafely a facility/location/crew is performing. Some theories point to the fact that a poor safety culture will inevitably create a horribly bad safety performance, while a good safety culture will produce an incident-free workplace.

Hogwash! First, there is no such thing as a safety culture. Second, a facility can have a great "safety culture" (i.e., everybody is turning in a good amount of safety observations and being "proactive," there hasn't been an incident in x number of days, people are friendly and welcoming, there is a good learning and mentoring environment, etc.) and still have an incident. I have recently worked at a facility that was known for having a "great safety-culture," in the words of the client rep from a large multinational corporation that we were working for. And then suddenly there were several recordable injuries back-to-back. Did we have a good safety culture or not? How was this safety culture assessed? Also, how could this have happened, seeing how we had such a great safety culture? Did the folks just wake up one day and decide, "You know what, I think I'll do something silly and get hurt today?" I will be digging a little deeper into this, but I am pretty sure you see where I am headed, or at least have a general idea.

How many times have you heard the term "finance culture, logistics culture, HR culture, cybersecurity culture," and so on within an organization? I am willing to bet never. So why are we so fascinated with a "safety culture?" At the end of the day, Safety/HSE is just another support mechanism or department within the greater enterprise umbrella, just the same as logistic, HR, finance, or cybersecurity. So why are we so determined to make something new and exciting out of a simple support element? To put simply – safety supports the operations or revenue generating arm of a company, and typically we cost money rather than generating revenue.

There are folks that will say, "I saved the company/companies X amount of money, etc., by improving the safety culture." Did you though? Or did you simply come in with an open mind, keen business acumen, and identify weaknesses within the safety support mechanism? Maybe you eliminated wastes within the safety support mechanism, but I bet the company is still spending money to keep that department running.

Back to safety culture, the one thing that we cannot agree upon a single definition of what it is or how we can tangibly measure it. It just feels right or some arbitrary number(s) are used to define what is good or bad. Again, there are folks, selling the prospect of how to do all this with their one shot, good-for-all-occasions products. If you find value in these scenarios, great! Keep working that system.

Here we are again referring to Pavlov and how we use arbitrary terms to define something that truly doesn't exist, and when those arbitrary terms are not met, we tell the workers that they aren't performing appropriately even though they may be under-budget, on schedule, the client is happy, and no one has been hurt doing it. Maybe we should start thinking of safety performance within the organizational culture and hopefully get the top brass truly bought in, spend the money to create impactful, meaningful, and worker-intuitive safety solutions for helping the workers work safer. I know, good luck with that, huh? I mean safety performance is the outcome of an organizational culture and not vice versa.

One last thing on safety culture before I move on. The wording "safety" culture implies that it is a "safety" thing which in turn takes ownership away from the workforce. It is a safety problem, therefore it's a problem for the safety guy to figure out, right? Terminology such as "organizational culture" brings ownership back to the organization as a whole, and more or less forces everyone from the top down to take ownership of and allow the HSE professionals to guide and mentor/coach, as it should be. If people do not feel as if they have control of something, they will typically either dismiss it or simply ride the waves because at the end of the day they have no control over the outcomes.

The same sentiment rings true with safety awards, where it's all or nothing. We have to remember that our employees can control their actions, and to some extent influence the actions of those around them. But that's it. They cannot control or influence the actions of someone working on the other side of the facility/site/plant. So, we need to get away from persecuting all for the mishaps of one or two and give ownership back to the employees.

This is truly the only way to see substantial change in organizational performance and culture. How about this—why don't we ask about the status of the organizational performance this quarter (i.e., safety, quality, budget, turnover, logistics, and schedule) and give rewards on the overall performance of the organization... . if we need a reason to quantify why we are showing the employees appreciation for showing up every day, working hard, and making the company money?

POPULAR LINGO

While we are on the topic of popular terminology, I also wanted to address things such as the latest and greatest advancements in the safety profession. How many times have you heard of a new SMS or that artificial intelligence will fix all your safety needs, or that a specific safety consulting company has this great new concept that is revolutionary and can increase your safety performance? First off, I don't want to necessarily discount anyone's knowledge or product but more of a generalized conversation. Sometimes, there are some good new concepts, products, services that come along, this is true. But many times, it is simply a revision of an old concept. The point? Just because it's novel doesn't necessarily make it better.

Recently I was reading through a popular professional networking site and came across a terrific graphic that sums this all up, some of you may have also seen it. There were two doors, one with a sign that read safety buzzwords with a long line of people waiting to get theirs (think Black Friday at a popular merchandiser). The other door had a sign that read actual safety, with one or two people standing in line. As the old saying goes, a picture can speak a thousand words, and there is so much truth in this simple graphic! Regardless of what buzzword you may or may not be using, the fact is that actual safety of the workforce lies with the workers and front-line supervisors, not in an office or in an SMS. The workers own their safety, safety professionals provide guidance, but we DO NOT "keep them safe." Worker safety is on the job site, identifying and mitigating risk, communicating, working together, and getting the job done without someone getting hurt, this is actual safety, and it is really that simple.

Unfortunately, all too often we try to complicate what actual safety is, for many variable reasons. Yes, we need an SMS to provide the guidance, but the worker must then appropriately utilize that guidance. Also, as unfortunate as it is, too many "safety professionals spend only a small amount of their time at the job sites. Without getting into a full-blown discussion of this, we must remember that safety happens at the work site and not in an office. As safety professionals, we need to be visiting the sites and having conversations with the crews at the very least. Let them know you are honestly interested in their concerns, either positive or negative. Be careful that you do not allow the employees to somehow be punished or receive punitive consequences for simply speaking freely with you in their feedback, comments, and concerns. When the employees feel empowered and take ownership of the site safety and feel as if there is someone who actually cares and can make an impactful change, this is that "secret silver bullet" and not some buzzword or new great concept. Sorry if you disagree and feel threatened by these comments, but it is the simple and actual truth. Human factors are the theme of this book and what is closer to the heart of this topic than actually making sure the workers feel empowered, and have bought into making their concerns heard and making the needed changes at the job site whenever they can?

While some of you may disagree with the ideals and topics I have discussed, this is perfectly fine, my hope was to generate thought and discourse. If your systems and processes are working for you, great! The general theme as you should have noticed by now is twofold. First, we must consider the person and how we should cater to the person, make the worksite easier for the person, and work diligently to anticipate where breakdowns and errors can occur. Second, ensure that ownership of safety and safe operations lies with the people actually doing and managing the work, not with someone in the office who is disconnected from the work activities.

HOP

Human and organizational performance (HOP) is one of the most recent theories/ methodologies that has gained considerable traction. While I am not here to advertise for one system or another, HOP seems to be most closely related to my way of thinking, at least to some degree. I have never been a big fan of placing a title on safety methodologies or systems. I believe safety is the ability to effectively and efficiently manage and mitigate errors rather than the more contemporary definition of safety being the absence of errors. This is why I say that I somewhat agree with the HOP methodology, namely because their first pillar is that errors will occur. This is a fact, and I have been making note of it from the first paragraph.

Another point of conversation is their pillar of learning from mistakes. I am a firm believer that people can and typically do learn from their mistakes. While there are many examples throughout the life of people repeating mistakes, consciously or unconsciously, I believe people do learn from their mistakes. If someone makes a critical mistake at work, generally they will learn from this mistake and readily move forward to becoming a better employee. It's happened to me. And let's be honest, just because someone makes a mistake does not make them a bad employee or a weak link in the organization. Because we all make mistakes, this line of thinking would make our entire organization consist of only weak links.

Have you ever heard something like "They were such a good employee; I can't believe they did that?" Just furthering my point, good people make mistakes, and we must be prepared for that in a positive and efficient manner, not just blame, shame, retrain, and hope for the best.

I'm not going to dive any further into HOP methodology or similarities between my beliefs and that system. I am one that does not believe we need a "system or methodology of safety" for people to work safely. As I already mentioned above, as safety professionals, we don't keep the employees safe; they (the workers) do. We provide the workers with tools, ideas, and guidance but they are the ones who are doing the work and, ultimately, they keep themselves safe. This is why it is crucial that the workers have ownership of their safety and it is not made to be a "safety thing" that lies in the realm of the safety guy/gal to deal with.

This brings me back to the "safety things are the safety manager's problem" type of mentality. I am referring to an organizational culture where workers, supervisors, and managers only see problems with safety-related issues and they believe ownership of these problems lies with the safety department, rather than viewing the problems as their own. A better way would be for these leaders to find solutions by utilizing the safety department as a subject matter expert (SME) that ensures the solutions are appropriate, meet industry and regulatory standards, etc.

In an ideal world with a mature organizational culture, workers/supervisors/managers take complete ownership of their problems because in the end the problem is their own to manage and rectify. All too often though we find the "safety things are the safety department's problem" and this mentality takes ownership and accountability away from the people doing and responsible for the work and places it on an ancillary support department. This ownership is a key element in having a proactive approach to safety performance, where the people responsible for the work are actively looking to resolve issues while utilizing SMEs, if and when further guidance or clarification is needed.

CONCLUSION

I know that I have touched on quite a few different topics, some more controversial than others, some of which may be seemingly common sense, and some that may seem potentially ridiculous to readers. My hope was to generate thought and bring the worker's perspective into the human factors discussion. All too many times when discussing various safety topics, we get hung up on the technical jargon and move away from the core, which is the people doing the work, which is unfortunate.

As always, compliance is an increasingly complex issue that must be included in all workplace health and safety discussions, whether it is internal, stakeholder, or governmental compliance, it is essential. But as I have covered, this is where we start to see the disconnect. When we are more concerned with compliance over feasibility, we have a problem. Guided by compliance, we often make rules/policies/procedures that just help us to be in compliance with some other rule/policy/procedure. This is not to say that compliance is not a concern but rather we need to ensure that our attempts at compliance are realistic to the work that is being undertaken and not just a paper exercise to meet some requirement, with solutions that are not workable in the field.

The other big talking point of this chapter was people and everyday life. As discussed, the people doing the work have a life outside of the job and we must be willing to understand and work with those daily concerns and problems the employees face in their daily life. This is the most basic root of human factors, all of the external factors that affect a person's work day, whether it is professional or personal, these events and factors mold how each individual employee will perform, both positively and negatively. To shun this idea is simply ignorant and to think that you can somehow create a policy to eliminate personal distractions from the work-place is naïve, at best.

All too many times it seems that we have gotten too far away from worrying about the employees' needs and lean too heavily on having a great system/SMS/book. Unfortunately, this will never work and is simply not a feasible approach. So, in closing, always remember that regardless of how good your "paper" is, you must consider how workable it is out where the money is made.

REFERENCES

Mcleod, S. (Feb, 2024). Pavlov's dogs experiment and Pavlovian conditioning response. www.simplypsychology.org/pavlov.html.
Vogt J, Leonhardt J, Koper B, Pennig S. (Feb 2023). Human factors in safety and business management. *Ergonomics*. 53(2):149–163. doi: 10.1080/00140130903248801. PMID: 20099171.

3 Good Leaders Want Safer Outcomes

Erich Pyles

INTRODUCTION

"Leadership is not just a theoretical concept to ponder but rather an applied discipline," as stated by Timothy R. Clark, a three-time CEO, Oxford-trained scholar, and consultant. According to Clark, leadership is the most crucial applied discipline globally. Any organization's prosperity, hardship, and misfortune are closely linked to leadership, which can be acquired through learning. However, many resources concentrate solely on skills, whereas true credibility and success stem from a blend of behavior, character, and competence.

To set the scene and clarify, I will not discuss the theoretical aspect of leadership in detail, but I want to highlight a couple of theories that set the stage for leaders, leadership, and behaviors that impact workers and the organizations they serve. Theories are not the be-all to leadership. Yet, they help us understand the foundational underpinning of relationships between leaders and workers and the overarching impact that leaders may have on an organization.

Furthermore, this chapter briefly analyzes leaders, leadership, behaviors, and their potential effects on employees and organizations. Drawing from my experience as a safety professional in the heavy industrial construction, oil and gas sectors, and as a practitioner and researcher of executive leadership, I have determined that safety leadership does not significantly differ from other forms of leadership. However, I will delve into the specifics of safety leadership, highlighting the key competencies that distinguish what I consider to be a foundational approach toward becoming an effective safety professional.

I have witnessed both the commendable and flawed aspects of leaders across different sectors, including those in the safety field. As a result, I have concluded that it would be more fitting to analyze leadership in a broader framework rather than solely focusing on the specific challenges and solutions related to safety leadership. However, I will briefly elaborate on what I believe establishes the foundation for effective safety leadership along with 10 competencies while building on the human-centric nature of our workforce.

Lastly, it is essential to acknowledge that leaders, their leadership styles, and their behavior can positively or negatively influence those around them. This impact

DOI: 10.1201/9781003583103-4

includes short-term effects and long-term consequences. Hence, this chapter delves into the psychosocial and psychological consequences and the long-lasting effects. Let's get started.

THEORIES IMPACTING LEADERSHIP

There are two theories that stand out when discussing leaders and followers: leader-member and social exchange theories. Leader-member exchange (LMX) relationships have been increasing for over 40 years. LMX theory suggests that leaders form high-quality social exchange relationships with some subordinates and low-quality economic exchange relationships with others. Peter Blau's social exchange theory (1964) is often utilized as the basis for studying leader-member exchange, where subordinates in social exchange relationships with their leaders tend to reciprocate with productive behaviors at work. However, the leader's behavior represents the other side of LMX. If a leader presents a negative behavior, workers will respond accordingly, impacting safety, quality, productivity, social exchange, and the overall work culture. I have seen both sides of the exchange and I know it is important to consider both sides.

As we drive down the road, walk into an operating facility, or step onto a construction project, some workers (boots on the ground) perform tedious, hands-on tasks that supervisors (leaders) have asked them to do. The workforce is the fundamental element of any organization, representing its most precious resource. I know that this may be hard to believe, but the combined efforts of employees directly influence the organization's overall success. As I have elaborated in discussions and meetings throughout my career, without boots on the ground or hands-on workers, individuals in supervisory positions would not have a job.

After my military (Army) career, I joined the local Boilermaker's Union (Local 667) out of Winfield, West Virginia. The union environment and work in and around power plants is where I started my construction journey. The hands-on approach and the valuable knowledge gained as a union member paved the path to where I am today as a safety professional. As a non-commissioned officer (senior enlisted), combat leader, and boilermaker, I did not know what LMX and social exchange theories were. However, from both career paths I took away good and bad leadership examples of how not to treat others and how I wanted my personnel to be treated. And I learned not to mistake the military leadership as requiring the same set of tools necessary for the Boilermakers. The leadership styles and mentality stretch from one end of the spectrum to the other. In other words, the leadership styles are entirely different.

I am sure you have encountered bad and good leaders who changed your mentality about leading. Leading is one thing, but the leader's behavior sets the tone for relationship and trust building. To have a strong relationship with our workers, we must learn to be servant leaders, providing performance support that supports the team and organizational culture and goals. There is a great passage in the Bible, found in the Gospel of Matthew, chapter 20, verses 25–28, where Jesus tells His disciples that effective leadership is not solely about exercising authority over people. Instead, the instruction continues, whoever wants to become great must lower himself to be a servant. People love servant leaders because they love the people they lead and because they view their leadership as service to the people they lead. Great leaders realize that serving others is the only way to lead with a pure heart, free of pride and arrogance.

Improving employee performance is equivalent to enhancing their abilities. The performance of employees serves as a catalyst for strengthening the organization's culture and overall achievements. The behavioral conduct and decisions made by leaders in the workplace play a pivotal role in shaping employee performance, acting as a bridge between the organization's objectives and strategic direction. Here is where the leader-member and social exchange theories come into play.

The absence of a conducive atmosphere indicates a deficiency in the dedication of leaders toward ensuring safety. In simpler terms, when leaders at every level are assertive and progressive and prioritize minimizing or eliminating risk, leading to safety, it becomes imperative for frontline supervisors and employees. The opposite is also true. When leaders do not take action addressing environmental health and safety (EHS) risks and technological challenges, such as insufficient funding for necessary resources or adequate staffing, they become directly responsible for the increased psychosocial hazards, errors, and psychological threats, ultimately culminating in unsafe behaviors.

From my experience and perspective, the role of leadership encompasses the ability to guide, coordinate, inspire, influence, respond, and envision to accomplish an organization's objectives and strategic aspirations. These proficiencies are indispensable when interacting with subordinates, superiors, colleagues in affiliated organizations, or the public. As an essential and crucial aspect of any effective leader, we must be capable of guiding others, inspiring them to dedicate their efforts and skills toward accomplishing the collective mission and objectives of the organization's strategic intent. As leaders, we cannot do it alone.

To enhance our effective leadership and influence skills, I find the following focused aspects to be crucial:

1. **Leading from within:** Understanding and harnessing our strengths, values, and beliefs to effectively guide and inspire others.
2. **Facilitating change:** Embracing change and actively supporting its implementation is essential for driving progress and achieving organizational goals.
3. **Building and sustaining trust:** Trust is the foundation of successful leadership. By consistently demonstrating integrity, transparency, and reliability, we can cultivate trust among our team members.
4. **Utilizing personal influence and political savvy:** Being aware of the dynamics of power and influence within an organization allows us to navigate complex situations and effectively advocate for our ideas and initiatives.
5. **Fostering an environment of leadership development:** Creating a culture that encourages continuous learning, growth, and mentorship enables individuals to develop their leadership potential and contribute to the organization's overall success.

6. **Promoting an environment of psychological safety:** By fostering an atmosphere where individuals feel safe to express their ideas, take risks, and learn from failures, we can encourage innovation, collaboration, and overall team effectiveness.

An individual operating outside the traditional leadership [status quo] model can pave the way for future change, express it as a vision, and inspire colleagues and themselves to question conventional thinking. Enhancing leadership effectiveness typically requires personal transformation, which may entail abandoning familiar habits and reevaluating unproductive behaviors. Let me warn you, however, that when stepping outside of the status quo (going along to get along) to promote or make positive changes, personally and professionally, you can be viewed as a disruptor.

A constant struggle faced by business, industry, and government leaders is the need to adjust and embrace the continuous and swift transformations that have become a regular occurrence in all facets of our lives. As leaders, it is essential that we acquire the skill of promptly and efficiently responding to change in order to thrive in the current business landscape. A successful leader can encourage others to accept and adapt to change. However, our behavior and communication style play a significant role in shaping the overall outcome. I have observed leaders at every level who are great at performing their jobs but not good at communicating or leading others. Numerous organizations, regardless of whether they are private or public, are currently grappling with successive waves of substantial employee turnover. In my experience, the construction industry, for many projects, sees an average turnover rate of 40%–50% or higher for the project's duration. The ongoing flux and the resulting consequences of ineffective leadership and employee retention challenges significantly impact both organizations and their projects.

We have heard the discussion for years: if we take care of the employees or frontline workers, they will take care of us, or any other similar saying. Yet, as leaders, we are continuing to fail at prioritization. Our priorities are still wrong after all these years. We are still seeing a mounting number of significant injuries and fatalities (SIFs) due to a lack of prioritization and effective leadership.

Yes, I meant to say that!

We should prioritize the health and welfare of the worker rather than placing productivity first. I am sure you have heard the following from leaders at every level – Safety (1st), Quality (2nd), then Productivity (3rd). I have listened to this for over a decade. Yet, many leaders are placing productivity first and safety last. Let's leave safety out of it for now. Before we can start discussing safety 1st, safety 2nd, safety 3rd, etc., we need to identify the hazards or risks associated with the task.

That said, we are not doing a good job of educating the workforce regarding identifying risk(s) for the task at hand. Many frontline leaders know what the risks are but fail to acknowledge how those risks may impact the worker and choose to look the other way for production's sake. Why is that? And we wonder why SIFs across all industry sectors continue to happen. Ultimately, the full extent of the mental and physical consequences caused by our inability to lead effectively may remain a mystery until it's too late.

LEADER AND LEADERSHIP FROM A PSYCHOLOGICAL PERSPECTIVE

The examination of leaders and leadership has been a prominent focus for decades. The concept of a "leader," previously linked mainly to historical figures such as Julius Caesar, Napoleon Bonaparte, Genghis Khan, Martin Luther King Jr., Rosa Parks, and General George S. Patton, has evolved to encompass political figures of diverse significance and is now utilized in organizational contexts to denote individuals who act as the central figure or facilitator of groups. While leadership positions were once viewed as a stroke of luck, contemporary society emphasizes recognizing and developing leadership skills to address the numerous situations where effective leadership is essential for organizational success.

Determining potential leaders' intellectual, personality, emotional, and relational traits is not an infallible science for organizational leaders. The selection process involves assessing various qualities, such as heightened assertiveness, strong interpersonal abilities, emotional intelligence, thirst for knowledge, exceptional listening skills, and the capacity to be empathetic and collaborate effectively within a team. Over the years, I have found that these attributes often separate the leaders from followers.

Not all leaders possess charismatic personalities, strong listening skills, or effective communication abilities. I will be the first to say that I have not always been charismatic, and I have had to develop my listening and communication skills over the years. Personality can be assessed by examining an individual's character and temperament, terms used to define a person's unique qualities. In the construction industry, leader personality traits vary widely; some are laid-back and approachable yet still manage to achieve their objectives by earning the trust of their workers and employing a people-oriented (human-centric) leadership style.

On the other hand, there are leaders who exhibit negative (destructive, toxic) traits, such as psychological abuse. These negative leaders focus solely on driving change and obtaining results (productivity-driven) without considering the human aspect, resulting in a stressful work environment that leads to employee mental fatigue, stress, burnout, and high turnover rates. Understanding how to empathize with others, prioritizing their perspectives over our ideas and beliefs, establishes us as a reliable source of support for those who seek our guidance, ultimately fostering trustworthy interpersonal connections.

As leaders at every level, it is of utmost importance to adopt a human-centric approach by actively listening, building trust, and effectively engaging with our personnel. To achieve this, we need to let go of any self-centered tendencies and prioritize the well-being of our team members. By temporarily setting aside our interests and preferences, we can focus on placing our personnel at the center of our attention. Additionally, it is essential to cultivate a non-judgmental mindset and discard any preconceived notions or stereotypes.

We as leaders must understand that the level of isolation we experience can directly impact our ability to maintain control over our leadership. Building and nurturing solid interpersonal relationships with the individuals we lead is essential in fostering a sense of connection and trust within the team. As leaders, we must recognize that every group includes individuals with unique characteristics, even if they share common goals. Therefore, we must tailor our approach to suit the diverse needs of our team members. This is particularly important in industries like construction, where working closely with small groups is common. Utilizing role-playing techniques can be beneficial in honing interpersonal skills through practical scenarios.

Leaders throughout the ages have been advised on the significance of setting an example and not expecting their team(s) to undertake tasks they themselves are unwilling to do. This approach, often referred to as leading from the front, may appear commendable, but it can be misguided and potentially hazardous. Some leaders exploit this notion as a justification to immerse themselves in their subordinates' work, engaging in minute details that are beyond the scope of their actual responsibilities and remuneration. Numerous leaders who are unwilling to delegate authority to their teams and are hesitant to step out of their comfort zones frequently rationalize these detrimental leadership shortcomings by claiming that they are leading by example.

Setting high standards for performance and behavior and being a role model of unwavering commitment is what it truly means to lead from the front. Demonstrating a strong work ethic, creative thinking, and a dedication to achieving tangible outcomes are essential components rather than simply going through the motions. Leading should be about showing our people how to stretch by constantly stretching ourselves, showing them how to be accountable by demonstrating our willingness to take accountability for outcomes. Leadership is about placing trust in people to do their jobs and giving them clear feedback on how they've performed.

As leaders, we are compensated to fulfill our responsibilities, and each level of leadership entails distinct tasks. Otherwise, the existence of hierarchical levels would be pointless if we were solely occupied with instructing workers on how to perform their duties. This type of leadership, known as "micromanaging," is detrimental to fostering trust among workers. Instead, micromanagement fosters a tense atmosphere, resulting in worker distrust, stress, and exhaustion. I have been micromanaged, and I am sure you can relate to or recall a micromanager. The feeling of facing a mental battle every day is not enjoyable.

THE PSYCHOSOCIAL RISKS AND IMPACTS IN THE WORKPLACE

In recent years, the work landscape has undergone significant changes due to various factors, including the implementation of advanced technology in work processes, increased workload and intensity, expansion in the construction industry, and evolving employment patterns – all of which I have been a part of at different levels. Consequently, these transformations have brought about new challenges that pose risks to employees' physical and mental well-being. In addition to the conventional hazards such as physical, biological, and chemical risks, there has been a rise in psychosocial risks resulting from these changes. Many of these risks, if not all, can be attributed to and controlled by leaders at every level.

Psychosocial hazards, job-related strain, aggression, and mistreatment are acknowledged as significant obstacles to occupational health and safety. Psychosocial hazards represent a prominent contemporary hurdle for ensuring occupational health and safety. That said, numerous research studies and books describe psychosocial hazards as having a negative impact on organizations. Yet, for many of us, we continue to see employees bearing and carrying the overwhelming weight of mental stress due to psychological and psychosocial hazards stemming from the workplace.

The International Labour Organization states that psychosocial hazards are connected to various factors that interact with one another. These factors encompass environmental and organizational conditions, such as job context, work organization, management, and employees' competencies and needs. Psychosocial hazards are pertinent to imbalances in the psychosocial realm and pertain to those interactions that demonstrate a detrimental impact on employees' health and mental well-being through their perceptions and experiences.

In simple terms, psychosocial hazards are defined as the combination of psychological and social factors. These hazards encompass various aspects of work design, the organization and management of work, as well as their social and environmental surroundings. They have the potential to cause harm to a person's psychological, social, or physical well-being.

Everyone agrees on what psychosocial hazards entail. These hazards cover things like company culture, job responsibilities, workload, relationships with coworkers and supervisors, and more. However, with constant changes in the workplace, new risks are emerging that we haven't even recognized yet.

Work experiences can tremendously impact an employee's psychological state of health. Research has shown that the impact of new working patterns and the risks that accompany them are apparent in employees' health and especially in their stress levels. According to the *Psychosocial Risk Management European Framework* (2008), continuous exposure to psychosocial risk factors may result in work-related stress for employees, affecting their efficiency in performing tasks. Occupational stress has garnered significant focus as it is now closely associated with work life. Approximately 40 million people, or nearly one in three workers in Europe, acknowledge being impacted by stress at work. In the United States, 83% of workers experience work-related stress, with a quarter of them identifying their job as the primary source of stress in their lives, as reported by Zippia. Inadequately handled workplace changes and lack of effective leadership and leader commitment can contribute to heightened stress levels and diminished job contentment, leading to employee burnout and turnover.

DESTRUCTIVE LEADERSHIP – PSYCHOLOGICAL AND PSYCHOSOCIAL IMPACT

Throughout recent decades, various terms in literature have been employed to elucidate the concept of destructive leadership, including abusive supervisors (Tepper,

2000), derailed leaders (Schackleton, 1995), or destructive leadership (Einarsen, Aasland & Skogstad, 2007). According to Padilla et al. (2007), destructive leaders have five essential traits: charisma, personalized power utilization, narcissism, negative life themes, and an ideology of hate. Therefore, this implies that destructive leadership could potentially stem from the personalities of the leaders themselves. From my experience, destructive leadership has often been linked to hostile attitudes or even psychopathological characteristics in the leader, not the work environment or the employee(s). As such, I would argue that leaders falter when they exhibit traits such as arrogance, vindictiveness, untrustworthiness, selfishness, emotional instability, compulsiveness, insensitivity, abrasiveness, or an inability to delegate or make decisions, and last but not least, they feel insecure in themselves and their abilities.

Throughout my life, in the military and construction, I have been exposed to leaders expressing good and bad narcissistic traits. The bad side of narcissism is what I am calling a personality trait characterized by arrogance, self-centeredness, entitlement, fragile self-worth, and hostility, which can lead to destructive patterns within the workplace. I believe that narcissistic leaders often have ambitious belief systems and leadership approaches, primarily driven by their desires for power, recognition, and admiration rather than genuine concern for their followers and organizations. If you are thinking that you know or have known leaders who fit this description, you are not alone. Most of us have.

It is undeniable that certain leaders possess personality traits that can lead to detrimental or even harmful behaviors in a professional setting. Yet, we must be able to step back, pause, and reflect before we fall into a destructive path, leading to a work environment producing psychological and psychosocial hazards that impact our workers. While conducting research, I found numerous studies (Chirico et al., 2019; Hon et al., 2023; Kortum, 2011) that widely recognized the significant influence of organizational leadership on employee's psychological and psychosocial safety and overall well-being. Moreover, research has shown that successful, genuine, and transformative leadership is linked to favorable health and safety results for employees. I have stepped into projects during a transition period where the project culture, incident rate (statistics), and employee turnover rates were due to verbally abusive, disruptive, and ineffective leadership. I am sure you have either heard of or been part of a project or organization where this happened. Situations of this nature not only negatively impact a project, leaders, and employees, but the actions or inactions of leadership can tarnish a company's name.

Effective leadership involves implementing a diverse set of top-notch leadership practices, and maintaining a consistent display of these behaviors can result in improved health and safety outcomes for employees in the workplace. There is no need to create a psychological (i.e., aggression, bullying, fatigue, harassment, and stress) or psychosocial (i.e., mental health, job demands, poor support, work demands, and traumatic events) hazardous work environment. Psychosocial hazards encompass various risks and dangers present in both work environments and society, capable of inflicting mental and physical harm. These hazards manifest as stress-inducing factors that can adversely affect an individual's physical, psychological, or overall well-being at work and home.

Workplace risks have distinct considerations for psychological and psychosocial safety. Effectively addressing psychosocial risks can significantly contribute to managing both psychological safety and hazards. Leaders and employees possessing knowledge or training in identifying symptoms of psychosocial risks can play a crucial role in mitigating psychological risks by intervening and reporting any issues. The term "psychosocial" refers to the connections between individuals and their thoughts, behaviors, and social surroundings. It encompasses the interaction between mental health or psychological aspects and social factors influencing an individual. Conversely, the term "psychological" refers to matters, emotions, and encounters associated with the human mind and mental well-being. Consequently, it can be inferred that psychological factors form a part of an individual's psychosocial condition, and effectively managing them can be highly advantageous.

Given the unique nature of each work environment and individual, there isn't a single method to ascertain whether something constitutes a psychosocial hazard. Nevertheless, according to the Institution of Occupational Safety and Health, a psychosocial hazard can be identified if it impacts any of the following:

- Work management
- Organizational structure
- Employee stress levels
- Worker health (physical, mental, or both)
- Employee turnover rate
- Job satisfaction

Many leaders can recognize a psychosocial hazard by observing how they and their employees respond to specific stressors and situations. Alternatively, these hazards can also be identified through:

1. **Company-wide surveys** – I would not target the whole organization. The result can (in most cases) be misleading from a holistic perspective. I recommend dividing the surveys into departments or teams if you want a better assessment of the overarching climate.
2. **Consultations** – Seek outside support to evaluate your project or organization by a third party that does not have a biased opinion.
3. **Injury reports** – Allow for comparisons with normative values to company, industry-specific, and national benchmarks while providing evidence that may provide answers to the human-centric or people-based quality aspects. In short, a deep dive into incidents may produce a more factual representation of why incidents are happening rather than taking incidents for face value.
4. **Absence reports** – These reports are, for the most part, never examined. Most organizations [leaders] never question an employee's absence. An employee's absence may be related to an employee's home life or workplace environment, where the absence of psychological needs is not being met.

5. **General workplace assessments** – If the assessments are to be effective, the individuals involved must be honest. Assessments are a snapshot of a moment in time and provide learning opportunities. For many, being truthful is problematic because it requires reflection.

In order to delve deeper into the topic and provide some clarification, I have provided the following examples that are likely the most prevalent instances of psychosocial hazards.

Work relationships: The absence of strong work relationships and solitary work environments can result in subpar interactions, ultimately impacting motivation. Conversely, incidents of bullying, discrimination, aggressive or violent conduct, and various forms of harassment can induce stress, thereby constituting psychosocial hazards.

Job insecurity and organizational changes: The absence of job security and potential organizational changes can lead to psychological strain, subsequently impacting an employee's physical and mental well-being. Additionally, the absence of acknowledgment for accomplishments may further exacerbate this situation.

Work demands: The demands of a specific role or assignment may exceed the capabilities of various individuals. Likewise, the absence of assistance or clear job expectations can pose a psychosocial risk, as experiencing isolation during times of stress can result in heightened stress levels and an unhealthy work–life balance, potentially resulting in burnout, injuries, and similar outcomes.

Traumatic events: Exposure to threatening and verbal abuse, potential harm (mentally and physically), distress, and fear resulting from violence in the workplace may result in anxiety, stress, trauma, and physical injury. Workplace violence can originate from internal sources within the organization or team, as well as external factors (i.e., clients) and so forth. These events (risks), if not controlled or eliminated, can negatively affect an employee's health, productivity, performance, and absenteeism. Therefore, this impacts the team, overall organizational goals, and strategic intent.

Psychosocial hazards have the potential to result in various adverse outcomes, including:

- Work-related stress
- Lack of motivation
- Injury
- Anxiety
- Fatigue
- Confusion
- Anger
- Depression
- Post-traumatic stress disorder
- Death

Psychosocial risks have the potential to impact not only the individual experiencing them but also those in their immediate social circle, such as family members, friends, and colleagues. The initial step in fostering employee health and safety within organizations involves extensively examining leadership styles and how they may impact various health and safety outcomes.

FACILITATING CHANGE

Implementing and overseeing change can be a challenging endeavor. Without a compelling and logical justification for initiating the change, it can be an uphill battle to garner the support and involvement of leaders at every level, let alone the employees. I have been with or part of organizations where change is needed to transform groups or teams to the next level. Yet, the communication surrounding the change(s) was unclear, leading to confusion at different levels. I am sure that you have either heard of or been part of the same situation. It is crucial for those involved to comprehend the following aspects:

1. The nature of the change
2. The reasons behind the necessity of the change
3. The implications of the change for both individuals and the organization
4. The urgency of implementing the change at this particular moment

I have listed specific qualities that we must possess to lead change effectively.

- Demonstrate trustworthiness, reliability, and influence.
- Introduce change without micromanaging or exerting control.
- Advocate for both organizational and individual needs by listening and supporting.
- Highlight the opportunities that come with change.
- Remain visible and accessible to the public throughout the process.
- Engaging individuals in change involves maintaining relationships with them throughout the transition. This phase entails:
  - Soliciting feedback throughout the process.
  - Acknowledging a variety of reactions to the change.
  - Responsively and genuinely addressing staff feedback and requests.

COMMUNICATE CHANGE

Communication serves as the foundational element that runs through all aspects of the change process. The ability to effectively convey information about change is crucial to a leader's ability to guide change initiatives successfully.

Many of us are familiar with the five Ws and one H approach (Who, What, When, Where, Why, and How) to crafting a news story. As leaders, we must adeptly communicate the narrative of change using a modified version of this structure. We should clarify:

- Who? Every individual involved in the process should understand their role in the change initiative: How their responsibilities will evolve, how they will contribute, and how they will shape the change process.
- What? What issue are we resolving with this change? What potential are we seeking to leverage? What will the organization look like?
- When? When will the changes need to take place or become effective?
- Where? Where do individuals fit in? Providing detailed information is key.
- Why? Why are the changes needed?
- How? Staff members need to grasp the unfolding of the plan. Plan unfolding entails outlining a step-by-step strategy for implementing the change and the timing and allocation of training and resources to support the transition.

It is essential for employees and stakeholders to comprehend the rationale behind the change and its alignment with the organization's overarching goals. As leaders, we must articulate the envisioned future. Granted, we may not always possess all the necessary details to address the five Ws and the H, it remains an obligation as a leader to gather the requisite information for those impacted by or involved in the change. Effective communication begins with actions rather than words. It is crucial to demonstrate consistency between our behavior and the values or vision we express. No matter how well-crafted our statements may be, they will only create confusion and distance if they contradict our actions.

In the realm of leadership, behavioral integrity is paramount when conveying change. As leaders, we are expected to embody the principles we advocate and act honestly. If we manipulate the truth or distort facts, people will lose trust in us, leading to a loss of respect. It is important to acknowledge that perceptions can become distorted during times of change. Employees will read into actions and statements, inferring underlying messages even when no explicit message is intended. To counter unfounded perceptions, it is crucial to keep our people informed and be transparent with them.

The "rule of six" should be kept in mind when communicating new information. People often need to receive information multiple times and through various channels before fully understanding it. Therefore, it is essential to communicate new information related to change at least six times and in different ways. Acknowledging the change or receipt of the changes is also critical. Additionally, it is crucial to anticipate and accommodate strong emotions during the change process. Allowing individuals to express their anger, frustration, confusion, anxiety, and other emotions in productive ways is necessary. Recognizing and communicating that emotions are a natural part of the transition process can help individuals navigate the change more effectively. For change to take place effectively in an organization requires mutual trust between the leader and those who are being encouraged to embrace change.

As leaders, we must possess proper character traits and be prepared with the appropriate and requisite knowledge. Then, we must turn our attention to the task at hand and take or enable action. Again, as leaders, we should be effective, establishing priorities and making decisive and confident decisions to accomplish tasks. Effective communication at all levels should be directed toward the attainment of team or organizational objectives.

There are two key factors that influence communication in relation to these goals. The first factor is language, which refers to the words used in leadership communication. The second factor is the leader's personality and tone when conveying the message to others.

Deborah J. Barrett (2010), an author and professor of professional communications at Rice University, emphasizes the importance of leaders effectively organizing simple and intricate correspondence. As leaders, it is crucial for us to communicate in a manner that ensures our intended audience comprehends the message. Furthermore, as leaders, we must be proficient in using language that is clear, accurate, and concise, enabling us to create and deliver confident and persuasive messages.

EFFECTIVE SAFETY LEADERSHIP AND COMPETENCIES

Safety leadership is paramount in shaping an organization's safety culture/climate. Safety leadership significantly impacts team members' mindset and behavior, risk awareness, compliance with safety protocols and regulations, and the overall work environment. Effective safety leadership establishes the culture's tone and directs teams toward their priorities and efforts.

Similar to other leadership roles, safety leadership can be categorized into two types: transactional and transformational. Transactional leadership primarily relies on procedures and regulations requiring a structured managerial framework. On the other hand, transformational leadership focuses on inspiring others to strive for excellence, necessitating extensive coordination, communication, and collaboration.

From my point of view, the distinction can be summarized in the following manner.

To attain proficient safety leadership, we must possess a combination of transactional and transformational leadership skills. The two skills ensure a balanced approach to safety leadership, encompassing both compliance-based practices (such as setting standards, monitoring employee performance, and providing feedback) and transformational leadership abilities (such as fostering a shared vision, collaborating to address safety concerns, engaging employees in a positive manner, and setting an example with effective safety practices).

The safety leadership competencies model that I am presenting outlines ten critical behaviors that in my opinion are essential for effective safety leadership performance.

1. **Supporting** team members by actively monitoring their decisions and actions, ensuring that they are aligned with the corporate strategy, vision, and values.
2. **Recognizing** and rewarding team members who demonstrate effective behavior.
3. **Engagement** of the workforce is needed for promotion and achievement.
4. **Actively caring** for team members' health, safety, and well-being.
5. **Collaborating** with team members and sharing ownership of safety by involving them in safety decision-making and empowering them to take personal responsibility for safety.

6. **Sharing a clear vision** for risk mitigation (risk tolerance) and facilitating the development of team goals, targets, and plans to achieve them.
7. **Inspiring** the team to strive for safety as the end result, not the first intention, through influential and encouraging communications.
8. **Leading by example** and exhibiting risk mitigation behaviors that set the standard for the team.
9. **Challenging** team members to think about risk awareness [harmful, hazardous] issues and scenarios in new and innovative ways.
10. **Empowering** team members and the workforce to bring forth concerns, issues, and risks that have the potential to impact the health and well-being of others, the project, or the organization.

The list may not represent all competencies that may be needed, and you may want to add to or create your list, but these ten competencies help support the foundation of effective safety leadership. They are crucial for creating a safe and secure work environment. Effective safety leadership increases discretionary effort and improves employee productivity, quality, and engagement. However, safety leaders who do not possess these skills may not prioritize the well-being of their employees or the overall goals of the organization's strategic intent. As a result, they may inadvertently create a toxic work environment characterized by blame and shame, which can be highly stressful for everyone involved.

LEADERSHIP THROUGH TRUST AND AUTHENTICITY

Have you ever considered what's at the core of highly successful leadership? There's a crucial element that is frequently overlooked in favor of strategic skills and charismatic communication – genuine (authenticity) and trust-based relationships. Exceptional leaders are those who cultivate trust, establishing an environment conducive to the growth of genuine connections. As a result, this enhances dedication, involvement, and productivity. Establishing trust goes beyond mere honesty or openness; it involves being dependable, consistent, and showing sincere concern for your team. Trust is not bestowed; it is acquired gradually, akin to the construction of a castle brick by brick.

As leaders, we can establish trust by adhering to our commitments. We must understand that actions carry more weight than mere words. Trust is nurtured when promises are fulfilled, and expectations are consistently met. Additionally, transparency plays a vital role in building trust. By maintaining open and honest communication, especially in challenging circumstances, as leaders, we must reassure our team members that they can rely on our guidance.

Furthermore, demonstrating respect extends beyond mere recognition of achievements; it encompasses appreciating the person's efforts and fostering a culture of psychological safety. Additionally, embracing feedback indicates a readiness to grow and evolve, showcasing genuine leadership qualities. Genuine connections are not only established within the confines of the workplace or meeting rooms. True relationships are formed through mutual experiences and engagements that reveal the personal side of those in leadership positions. Dedicate time to comprehend the world of your team members, including their obstacles and successes. It is in these instances of mutual openness that authenticity truly stands out.

To conclude, bear in mind the profound advice of Simon Sinek: A team does not merely consist of individuals who collaborate, but rather it comprises individuals who have faith in one another. As a leader, your capacity to cultivate trust and establish genuine connections is not merely a commendable practice, but rather it is the fundamental core of effective leadership.

CONCLUSION

It is widely agreed upon in the extensive body of leadership literature that leadership encompasses far more than just positive actions. Leadership is a constantly evolving process that involves collaboration between leaders, followers, and the surrounding environment. The outcome of the leadership process significantly impacts both group and organizational achievements. Leadership is about guiding and influencing outcomes, enabling groups of people to work together toward common goals. It's not limited to executives or managers; anyone can exhibit leadership qualities.

A well-known author and public speaker, Myles Munroe, famously stated, "The most devastating tragedy in one's existence is not the end of life, but rather a life devoid of meaning and purpose." The reason behind our actions provides significance and guidance in our lives. It motivates us to move ahead, molds our choices, and impacts our behavior. Understanding our purpose allows us to synchronize our actions with a higher cause. Leaders who are driven by purpose influence their team members and establish a collective vision and dedication in their companies.

Although there is a comprehensive understanding of the importance of "great" leadership, there is also a growing emphasis on the study of "destructive leadership." I brought forth the attributes, behaviors, and traits of leaders, their leadership approach, and their subsequent effects on the employees and organizations they represent. With this information, it is up to you to decide who you want to be, how others view you, and how you want to lead.

Developing leadership skills cannot be achieved through a brief workshop or a short training session provided by your organization. It requires significant time, dedication, and consistent practice over months or years. According to Simon Goncharenko in his book, *Save Lives: Pushing Boundaries in Human Factors,* "good leaders are easy to follow, as they inspire, encourage, and empower their teams to succeed." As individuals in leadership positions, it is crucial to prioritize the mastery of leadership communication if we aspire to lead others in any industry or field effectively. By doing so, we can inspire trust, gain followership [relationships], and establish ourselves as a respected leader within our team(s), organization, or the wider community.

In summary, purpose-driven leadership involves understanding our purpose, serving others, and leading authentically. It requires self-discipline and a commitment to making a positive impact. Keep in mind that leadership is not about being the most intelligent individual in the room; it is about empowering others and striving to make our organizations and the world a better place for those around us. Ultimately, the actions we take are more significant than the information we consume. In order to effectively lead, our words should align with our beliefs and actions. It is through our actions that we truly make an impact.

BIBLIOGRAPHY

Barrett, D. (2010). *Leadership communication*. McGraw-Hill/Irwin.

Chirico, F., Heponiemi, T., Pavlova, M., Zaffina, S., & Magnavita, N. (2019). Psychosocial risk prevention in a global occupational health perspective. A descriptive analysis. *International Journal of Environmental Research and Public Health*, 16(14), 2470.

Clark, T. R. (2016). *Leading with character and competence: Moving beyond title, position, and authority*. Berrett-Koehler Publishers.

Einarsen, S., Aasland, M., & Skogstad, A. (2007). Destructive leadership behaviour: A definition and conceptual model. *The Leadership Quarterly*, 18(3), 207–216. https://doi.org/10.1016/j.leaqua.2007.03.002

Goncharenko, Simon (2024). *Save lives: Pushing boundaries in human factors*. Kindle Press.

Hon, C. K., Sun, C., Way, K. A., Jimmieson, N. L., Xia, B., & Biggs, H. C. (2023). *Psychosocial hazards affecting mental health in the construction industry: A qualitative study in Australia*. Engineering, Construction and Architectural Management.

Kortum, E. (2011). Perceptions of psychosocial hazards, work-related stress and workplace priority risks in developing countries. *Journal of Occupational Health*, 53(2), 144–155.

Leka, S., Griffiths, A., & Cox, T. (2003). *Work Organization and Stress*. World Health Organization.

Padilla, A., Hagan, R., & Kaiser, R. B. (2007). The toxic triangle: Destructive leaders, susceptible followers, and conducive environments. *The Leadership Quarterly*, 18(3), 179–194.

Shackleton, V. (1995). Leaders who derail. *Business Leadership*, pp. 89–100.

Tepper, B. J. (2000). Consequences of abusive supervision. *Academy of Management Journal*, 43(2), 178–190. www.jstor.org/stable/1556375

4 Balancing Progressive Discipline with a No-Blame Culture in Workplace Safety

James A. Junkin

INTRODUCTION

Creating a safe and productive work environment involves a challenging task for organizations: striking a balance between progressive discipline policies and fostering a no-blame culture. These two elements, while seemingly at odds, play crucial roles in maintaining workplace safety. Integrating them effectively requires a strategic approach that ensures they support rather than contradict each other. In this chapter, we will explore how to harmonize these policies, offering practical insights and examples to illustrate their integration.

As the chief executive officer of Mariner-Gulf Consulting & Services, LLC, the chair of the Veriforce Strategic Advisory Board, and the past chair of *Professional Safety Journal's* editorial review board, I have advised many corporations around the world, investigated 16 workplace fatalities, audited hundreds of contractors, including for one of the top five global tech giants, and assisted with numerous regulatory interactions, including BP's Horizon incident in the Gulf of Mexico. Over the course of my professional experience and thousands of interactions that it afforded me, I have made some helpful observations about organizational culture, which are shared in this chapter.

UNDERSTANDING PROGRESSIVE DISCIPLINE POLICIES

Progressive discipline policies are structured to address and correct undesirable employee behaviors through a series of escalating actions. Starting with verbal warnings and escalating to written warnings, suspensions, and, ultimately, termination, the goal is not only to maintain a productive workplace but also to give employees the opportunity to rectify their behavior before harsher consequences are imposed.

DOI: 10.1201/9781003583103-5

PURPOSES OF PROGRESSIVE DISCIPLINE

Progressive discipline serves several important functions:

- **Consistency**: It ensures that all employees are treated fairly, with the same standards applied across the organization.
- **Documentation**: These steps create a formal record of efforts to address behavioral issues, which can be critical in legal disputes.
- **Opportunity for Improvement**: Progressive discipline provides clear guidelines, allowing employees a chance to modify their behavior before facing severe consequences.

THE NO-BLAME CULTURE IN SAFETY

In contrast to traditional discipline, a no-blame culture emphasizes identifying and addressing the root causes of safety incidents, focusing on systems and processes rather than individual faults. The underlying belief is that most incidents are due to systemic failures, such as insufficient training or flawed procedures, rather than individual negligence.

KEY PRINCIPLES OF A NO-BLAME CULTURE

A no-blame culture rests on several foundational principles:

- **Root Cause Analysis**: Investigating incidents to identify systemic failures rather than focusing on individual responsibility.
- **Continuous Improvement**: Using safety incidents as opportunities to learn and improve processes.
- **Psychological Safety**: Creating an environment where employees feel safe to report mistakes and near-misses without fear of punishment.
- **Transparency**: Encouraging open communication around safety issues.
- **Collaboration**: Promoting teamwork and shared responsibility for safety.

BENEFITS OF A NO-BLAME CULTURE

Adopting a no-blame culture brings significant advantages to workplace safety:

- Employees are more likely to report safety issues without fear of repercussions.
- A focus on systemic issues leads to more effective investigations and prevention.
- Employees feel valued and become more actively involved in safety efforts.
- By removing the fear of blame, the workplace becomes more positive and supportive.
- The organization continually learns from incidents, fostering ongoing improvements.

IMPLEMENTATION STRATEGIES FOR A NO-BLAME CULTURE

To establish a no-blame culture, organizations need to adopt specific strategies:

- Leaders must be visible champions of the no-blame approach.
- Provide employees with comprehensive training on the value and principles of a no-blame culture.
- Ensure that expectations around safety and reporting are clearly communicated.
- Use systematic investigations to uncover the real causes of safety incidents.
- Encourage and reward employees who actively report safety issues.
- Regularly assess and adjust the implementation of the no-blame culture as needed.

BALANCING THE TWO APPROACHES

The key to successfully blending progressive discipline with a no-blame culture lies in the distinction between behavior and systemic issues. Some incidents result from system failures, while others stem from conscious choices to ignore safety rules. Balancing the two approaches requires careful navigation.

STRATEGIES FOR INTEGRATION

As you navigate the waters of integrating the progressive discipline within the no-blame culture, consider the following as starting points:

- **Separate Behavior from Systems Issues**: Recognize when incidents are caused by flaws in processes and when they are due to deliberate actions.
- **Just Culture Framework**: Adopt a "just culture" approach that differentiates between genuine mistakes, reckless behavior, and willful violations. Employees are held accountable only for deliberate breaches of safety protocols. And even when a deliberate breach is established, it is important to probe it further, in order to test whether culture, external pressures, unwritten expectations, unclear instructions, conflicting procedures, or anything else played a role.
- **Transparent Communication**: Maintain open lines of communication, so employees understand the difference between situations that lead to disciplinary action and those that result in learning opportunities.
- **Emphasize Training and Support**: When gaps in knowledge are identified, provide training and support rather than resorting to discipline.
- **Encourage Reporting**: Promote a reporting culture where employees feel safe to share their concerns without fear of retribution. Along with the culture, ensure that your reporting channels are intuitive, easy to use, and are fully operational with feedback on the back end, as there is nothing

that discourages reporting more than inadequate or convoluted methods for doing so.

- **Continuous Review**: Regularly assess both progressive discipline and no-blame practices, adjusting them to ensure they remain effective.

CASE STUDIES

Perhaps some examples of how this works in real life may be helpful for driving the point further.

Example 1: Near-Miss Reporting and Root Cause Analysis
An employee reports a near-miss involving a fall from a ladder due to a faulty rung. Instead of assigning blame, the organization conducted a root cause analysis, discovering that maintenance procedures were inadequate. As a result, the maintenance team is retrained, and the employee is encouraged to continue reporting near-misses. This approach identifies systemic problems and fosters a culture of continuous improvement.

Example 2: Just Culture Framework in Action
An employee operating a forklift causes minor damage to equipment. Using a just culture framework, the organization determines that the employee's actions were not reckless. Instead of discipline, the employee receives additional coaching on safe forklift operation, reinforcing the importance of safety without punishment.

Example 3: Addressing Behavioral Issues
An employee repeatedly fails to wear personal protective equipment (PPE). After ensuring that there are no fit issues, it was determined that this is a behavioral concern and the organization delivers a formal warning and provides additional training on the importance of PPE. This approach holds the employee accountable without attributing the failure to broader systemic issues.

JUST PROGRESSIVE DISCIPLINE: ACCOUNTABILITY WITHOUT FEAR

While a no-blame culture is essential for fostering openness and communication, there are situations where accountability is necessary. This is where just progressive discipline comes into play. A just culture balances a no-blame approach with accountability, especially in cases of willful violations or reckless behavior. The goal is not punishment for its own sake but to ensure that everyone in the organization is committed to maintaining a safe environment.

At the heart of any successful safety program is reporting—the process by which employees communicate incidents, accidents, and hazards to management. Effective reporting affords the organization the ability to learn and be nimble in adjusting its approach to risk and communicating the lessons learned across its locations. The effectiveness of reporting systems depends on how safe employees feel when raising concerns and the trust they have in the organization to respond appropriately.

ENCOURAGING INCIDENT AND HAZARD REPORTING THROUGH NO-BLAME CULTURE

A no-blame culture encourages employees to report hazards and unsafe conditions by:

1. **Promoting Psychological Safety**: Employees are less likely to hide mistakes or unsafe conditions when they feel psychologically safe. In a no-blame culture, workers know that reporting a hazard will lead to improvements, not reprimands.
   a. **Example**: In a manufacturing plant, an employee notices a poorly maintained machine that could cause injury. Instead of fearing punishment for not noticing sooner, they report the hazard. The machine is repaired, and the employee is recognized for their contribution to safety.
2. **Improving Data Collection**: With more open reporting, organizations can collect a wealth of data on near-misses, minor incidents, and hazards. This data provides valuable insights into trends and areas of concern, allowing management to implement preventive measures before an accident occurs.
3. **Creating a Learning Environment**: When workers report incidents, they contribute to a culture of learning. Instead of focusing on blame, management can use the information to assess processes, provide additional training, or modify procedures to enhance safety.
   a. **Example**: An employee reports a near-miss involving forklift operation. Instead of discipline, the incident is analyzed, leading to improved training on forklift safety for all operators.

REINFORCING SAFETY THROUGH JUST PROGRESSIVE DISCIPLINE

While a no-blame culture promotes openness, just progressive discipline ensures that accountability remains in place for actions that could endanger others. It serves as a safeguard against reckless behavior, reinforcing that safety rules must be followed.

1. **Reinforcing Safety Protocols**: By applying discipline only in cases of willful or reckless violations, just progressive discipline reinforces the importance of following safety procedures. Employees understand that while mistakes are opportunities to learn, deliberate disregard for safety will not be tolerated.
   a. **Example**: An employee repeatedly fails to wear PPE despite multiple warnings. Progressive discipline is applied, starting with a verbal warning, escalating to a written warning, and eventually suspension if the behavior persists. This sends a clear message that while reporting hazards is encouraged, willful safety violations have consequences.
2. **Establishing Trust and Fairness**: Employees are more likely to report hazards if they trust that the organization will act fairly. Just progressive discipline builds this trust by ensuring that discipline is applied consistently but only when truly warranted.
3. **Deterring Reckless Behavior**: Just progressive discipline acts as a deterrent to reckless or negligent actions, which can put the entire workplace at risk.

By holding employees accountable for dangerous behaviors, organizations can maintain high safety standards while still encouraging honest reporting of mistakes and hazards.

CASE STUDIES: ENHANCING REPORTING AND SAFETY THROUGH BALANCED APPROACHES

Though the aforementioned content was delivered in a purposefully easy-to-digest format, people always learn better when stories reinforce the main points. So, the following case studies should drive the points home and clarify any misunderstandings.

CASE STUDY 1: REPORTING NEAR-MISSES IN A CONSTRUCTION COMPANY

In a large construction company, near-misses were often ignored due to a fear of blame. Management implemented a no-blame culture, emphasizing that all reports would be used for learning and improvement. Over time, the number of reported near-misses skyrocketed. For example, a worker reported a near-miss involving a scaffold that had been improperly assembled. The report led to a comprehensive review of scaffold assembly procedures and training, preventing future incidents.

CASE STUDY 2: ADDRESSING RECKLESS BEHAVIOR IN A CHEMICAL PLANT

In a chemical plant, safety rules were strictly enforced, but incidents involving PPE were still frequent. Management introduced a just progressive discipline system alongside their no-blame culture. While most incidents were addressed through root cause analysis and training, deliberate non-compliance with PPE use resulted in progressive discipline. This dual approach reduced PPE violations significantly while maintaining a culture of openness in reporting other hazards.

CONCLUSION: CREATING A REPORTING CULTURE FOR WORKPLACE SAFETY

The combination of a no-blame culture and just progressive discipline is essential in creating an environment where incident, accident, and hazard reporting thrives. By removing the fear of blame and punishment, employees are encouraged to contribute to workplace safety proactively. At the same time, just progressive discipline ensures that accountability is maintained, and deliberate violations of safety protocols are addressed.

When these two approaches work in harmony, organizations experience enhanced safety, enhanced incident reporting, and a culture of continuous improvement. The ultimate outcome is not just compliance with safety regulations but the proactive identification and correction of unsafe conditions, leading to fewer accidents and a safer, more productive workplace.

REFERENCES

Albert, A., Pandit, B., Patil, Y., & Louis, J. (2020). Does the potential safety risk affect whether particular construction hazards are recognized or not?. *Journal of Safety Research, 75,* 241–250.

Dekker, S. (2016). *Just culture: Balancing safety and accountability.* CRC Press.

Koolwijk, J. S. J., van Oel, C. J., & Gaviria Moreno, J. C. (2020). No-blame culture and the effectiveness of project-based design teams in the construction industry: The mediating role of teamwork. *Journal of Management in Engineering, 36*(4), 04020033.

Lloyd-Walker, B. M., Mills, A. J., & Walker, D. H. (2014). Enabling construction innovation: The role of a no-blame culture as a collaboration behavioural driver in project alliances. *Construction Management and Economics, 32*(3), 229–245.

Lupton, B., & Warren, R. (2018). Managing without blame? Insights from the philosophy of blame. *Journal of Business Ethics, 152,* 41–52.

McCall, J. R., & Pruchnicki, S. (2017). Just culture: A case study of accountability relationship boundaries influence on safety in HIGH-consequence industries. *Safety Science, 94,* 143–151.

Obenauer, W. G., & Kalsher, M. J. (2023). Does blame always shift? Examining the impact of workplace safety communication language on post-accident blame attributions for multiple entities. *Acta Psychologica, 240,* 104024.

Parker, J., & Davies, B. (2020). No blame no gain? From a no blame culture to a responsibility culture in medicine. *Journal of Applied Philosophy, 37*(4), 646–660.

Petitta, L., Probst, T. M., & Barbaranelli, C. (2017). Safety culture, moral disengagement, and accident underreporting. *Journal of Business Ethics, 141,* 489–504.

Rami, U., & Gould, C. (2016). From a "culture of blame" to an encouraged "learning from failure culture". *Business Perspectives and Research, 4*(2), 161–168.

Sherratt, F., Thallapureddy, S., Bhandari, S., Hansen, H., Harch, D., & Hallowell, M. R. (2023). The unintended consequences of no blame ideology for incident investigation in the US construction industry. *Safety Science, 166,* 106247.

Smith, S. D., Sherratt, F., & Oswald, D. C. (2017). The antecedents and development of unsafety. *Proceedings of the Institution of Civil Engineers-Management, Procurement and Law, 170*(2), 59–67.

Stefana, E., De Paola, E., Snaiderbaur Bono, C. S., Bianchini, F., Vagheggi, T., & Patriarca, R. (2024). Beyond blame: A systemic accident analysis through a neutralized human factors taxonomy. *Human Factors and Ergonomics in Manufacturing, 34*(5).

Thallapureddy, S., Sherratt, F., Bhandari, S., Hallowell, M., & Hansen, H. (2023). Exploring bias in incident investigations: An empirical examination using construction case studies. *Journal of Safety Research, 86,* 336–345.

van Marrewijk, A., & van der Steen, H. (2024). Organizational learning from construction fatalities: Balancing juridical, ethical, and operational processes. *Safety Science, 174,* 106472.

Walton, M. (2004). Creating a "no blame" culture: have we got the balance right?. *BMJ Quality & Safety, 13*(3), 163–164.

5 The Role of AI in Human-Centric Risk Decision-Making

Steven Haynes

INTRODUCTION

The success of a safety program can be reduced to just a couple of factors, but one factor undoubtedly plays a more significant role than others: effective decision-making. Safety professionals face high-stake environments where every decision can significantly impact the well-being of individuals, operational efficiency, and compliance. The complexities involved in managing risk have grown exponentially with the rapid pace of industry change, often characterized by complex regulatory demands, heightened public scrutiny, and unforeseen external risks. Consequently, decision-making in this field is not simply about following procedural steps; it requires a dynamic approach that balances quantitative analysis with qualitative judgment, factoring in unique operational conditions and potential safety outcomes.

At the same time, emerging technologies, especially automation and artificial intelligence (AI), are reshaping the methods available for risk assessment and decision-making. Safety professionals increasingly seek to use predictive analytics, automated monitoring, and advanced data analysis platforms. In comparison, traditional risk management relies heavily on human analysis and experience-based intuition. AI and machine learning (ML) offer new avenues for assessing patterns in large datasets and identifying risks in real time. A task that might have taken weeks to complete by a human can now often be executed by AI in minutes or even seconds. These technologies can flag unusual patterns, perform real-time monitoring, and even predict potential incidents before they occur, providing insights that would be otherwise inaccessible.

Integrating automation and AI into decision-making brings new challenges but with apparent advantages. Questions about overreliance on technology, the potential for automation bias, and the ethical implications of AI in safety decisions are pressing concerns. While these tools are highly effective in processing information at scale, they lack the nuanced understanding and contextual insights that human decision-makers bring. Safety professionals must, therefore, consider the limitations of AI when making final decisions, ensuring that the technology serves as a support system rather than as a replacement.

DOI: 10.1201/9781003583103-6

Throughout this chapter, safety professionals will be introduced to a comprehensive understanding of risk decision-making by exploring the traditional models and frameworks guiding the field while highlighting the expanding role of automation and AI. By understanding the foundational principles of decision-making and the opportunities presented by new technologies, safety professionals can make more informed, balanced decisions. Integrating human expertise with advanced technological tools will enable more resilient, adaptive safety practices to address current and future risks. Through this lens, safety professionals will be better positioned to navigate the evolving risk landscape, balancing technological insights with their essential human judgment to protect both individuals and organizations.

FOUNDATIONS OF DECISION-MAKING IN RISK MANAGEMENT

Effective decision-making is crucial for safety professionals working within complex, high-stake environments. The primary focus of risk management is to facilitate decisions that minimize potential harm to individuals and assets, ensure regulatory compliance, and promote operational resilience. Safety professionals must be well-versed in several core decision-making frameworks and approaches to accomplish this. The following section examines traditional decision-making models, reviews cognitive biases that impact decisions, and explores ways to mitigate bias, setting the stage for integrating advanced tools like AI to enhance these foundational techniques.

Traditional decision-making models provide structured approaches to evaluate and select the best course of action in risk scenarios. Two commonly used models are the Rational Decision-Making Model and Herbert Simon's Bounded Rationality Model (Kalantari, 2010).

RATIONAL DECISION-MAKING MODEL

The Rational Decision-Making Model is rooted in a systematic, step-by-step process that includes defining the problem, gathering data, identifying possible solutions, and weighing each option's costs and benefits (Simon, 1979). This model assumes that decision-makers have access to all relevant information and can analyze it objectively. Regarding safety decisions, the rational model helps identify clear, data-driven solutions to mitigate risks, creating an ordered, evidence-based approach to preventing accidents.

However, the rational model's limitations become obvious in dynamic safety contexts where risks evolve rapidly. While safety professionals aim to make decisions based on empirical evidence, time constraints, incomplete data, and rapidly changing environments make it challenging to apply this model strictly. When time is not on the side of the safety professional, a more organic approach will likely be required—hence the Bounded Rationality Model.

BOUNDED RATIONALITY MODEL

The Bounded Rationality Model, proposed by Herbert Simon (1955), recognizes that decision-makers often work under limited information, time pressures, and cognitive limitations. Simon suggested that instead of finding the "optimal" solution, decision-makers frequently settle for a solution that is "good enough"—a process known as satisficing (Ben-Haim, 2012; Schwartz et al., 2011). Another way of looking at this model is that there is never enough time and information to make the perfect decision, so good enough is what we will likely follow.

This model is highly applicable in safety management. In time-sensitive situations where information is limited or ambiguous, bounded rationality helps safety professionals act quickly based on the best available information rather than delaying action in search of the perfect solution. For example, waiting for comprehensive data could hinder effective action during an emergency response. The bounded rationality model supports decision-makers in making prompt, albeit imperfect, decisions that aim to mitigate risk effectively.

To put it bluntly, safety professionals must make the right decision at the right time for the right reasons.

INTUITIVE VS. ANALYTICAL APPROACHES TO SAFETY DECISIONS

Safety professionals often toggle between intuitive and analytical approaches when making risk-related decisions. Analytical decision-making involves systematically analyzing data and weighing evidence, aligning with traditional empirical models. However, intuitive decision-making, which relies on experience and instinct, is also a significant factor, especially in high-pressure situations where analytical methods might be too slow, or the data is lacking. The strengths and weaknesses of the two approaches are discussed here.

Analytical approaches are beneficial for assessing complex situations that involve multiple variables and high levels of uncertainty. These approaches allow safety professionals to incorporate quantitative data into their decision-making process, such as past incident reports, near-miss reports, previous loss run (claims) reports, and risk assessments. This data-driven focus can help pinpoint root causes of risk, evaluate potential safety interventions, and prioritize actions based on the most likely or severe impact. An obvious limitation of this approach is emerging risks, where data is unavailable, the circumstances are novel, and past experiences do not reflect current conditions.

Intuitive decision-making is based on pattern recognition and accumulated experience, which can be invaluable in real-time crises. Professionals with extensive industry experience often possess "gut instincts" developed over years of handling various safety scenarios. While intuition should not replace analytical methods, it can complement them, especially in situations that demand immediate action. Studies show that experienced professionals often make quicker and more accurate decisions in high-stakes situations due to their ability to identify patterns subconsciously (Bilalić et al., 2010; Whittaker, 2018). As it is intuitive, this approach may produce more creative solutions, but greater organizational flexibility is needed to utilize it effectively. A top-down approach from highly mechanistic organizations would not likely yield the intended results.

A balance between these two approaches is necessary to make the optimal decision during normal and abnormal times, but safety professionals will likely favor one approach over another based on their circumstances. As a result, all decision-makers must be aware of their biases to facilitate the right decision for the right reason.

Cognitive Biases and Heuristics in Decision-Making

While intuitive decision-making has benefits, it also introduces the risk of cognitive biases, which can lead to errors in judgment. Safety professionals must be aware of these biases, as they can compromise the objectivity and accuracy of their decisions. Some of the most common biases that impact decision-making in risk management include the following:

Confirmation Bias occurs when decision-makers focus on information confirming their beliefs while disregarding contradictory evidence (Cassell et al., 2022). For example, a safety professional might emphasize past data supporting a particular intervention while ignoring emerging evidence suggesting a new hazard. To reduce the effect of confirmation bias, decision-makers should seek out and weigh opposing viewpoints, encourage diverse perspectives, and incorporate data from multiple sources.

Anchoring Bias is the tendency to rely heavily on the first piece of information received when making a decision, even if it is not the most relevant (Lieder et al., 2018). In a safety context, initial data about a potential hazard could unduly influence a professional's risk assessment, potentially leading to an inaccurate evaluation of the threat. Regularly reassessing initial assumptions and updating them with new information is essential for overcoming anchoring bias.

Availability Heuristic is when individuals overestimate the likelihood of an event based on how easily examples come to mind (Folkes, 1988). If a recent incident of a specific risk (e.g., a fire) is fresh in memory, a safety professional might give disproportionate weight to fire risks even when other hazards, such as slips, trips, and falls, are statistically more probable. Risk assessments should be based on comprehensive data rather than recent experiences to counteract this heuristic.

Status Quo Bias is a preference for the current state of affairs, leading professionals to resist changes even if they could reduce risk (Samuelson & Zeckhauser, 1988). In safety management, this bias can hinder the adoption of new safety practices or technologies. Overcoming status quo bias requires fostering a culture of continuous improvement and emphasizing the potential benefits of new approaches or technologies.

There are dozens of other types of biases that can hinder decision-making, but these four biases are likely biases that safety professionals can relate to. All safety professionals would benefit from training on heuristics and biases for themselves and their workforce, as biases often lead to poor organizational safety culture and policy administration.

TECHNIQUES FOR MITIGATING BIAS IN SAFETY SETTINGS

Safety professionals can implement strategies that reduce cognitive biases to improve the quality of risk-related decisions. Some effective techniques include:

Structured Decision-Making Tools: Using structured decision-making tools like decision trees, checklists, and risk matrices can help professionals maintain an objective focus by ensuring that all relevant factors are considered. These tools guide decision-makers through a standardized process, reducing the likelihood of biased judgments.

Peer Review and Collaboration: Involving multiple perspectives can help offset individual biases and promote more balanced decisions. Peer reviews and collaborative decision-making foster a diversity of thought, which can be especially beneficial when assessing complex risks. One common factor this researcher noticed in organizations suffering from poor safety culture is the lack of diversity in the overall safety program.

Training and Awareness Programs: Educating safety professionals about common biases and heuristics can increase self-awareness and encourage vigilance against biased thinking. Training programs on decision-making under uncertainty can improve professionals' ability to recognize and counteract biases in real-world situations. Today's training programs are often delivered asynchronously and online, depriving employees of getting to know others in their organization. Regarding safety training, promoting in-person and hands-on training will likely create more resiliency during high-risk situations, as people are familiar with one another.

Scenario Analysis and Pre-Event Exercises: Scenario analysis and pre-event exercises allow teams to anticipate potential failures before they occur, prompting a proactive approach to risk mitigation. By imagining a scenario where a safety plan fails, professionals are encouraged to identify overlooked risks and adjust their strategy accordingly.

This foundation on decision-making prepares safety professionals to evaluate complex risk scenarios using a mix of rational analysis, practical heuristics, and collaborative strategies, positioning them to incorporate advanced technologies in subsequent steps effectively.

FRAMEWORKS FOR EFFECTIVE SAFETY DECISION-MAKING

To make informed decisions in safety management, professionals rely on structured frameworks that allow for systematic assessment, prioritization, and action. These frameworks ensure that all relevant factors are considered, risks are accurately assessed, and decisions are made to protect people, assets, and regulatory compliance. This section discusses quantitative, qualitative, and hybrid approaches that equip safety professionals with the tools necessary to navigate complex risk scenarios and make decisions grounded in data and professional judgment.

QUANTITATIVE APPROACHES

Quantitative approaches to risk management rely on numerical data and statistical analysis to evaluate potential risks and weigh the costs and benefits of different safety interventions. In safety management, quantitative methods can be lifesaving in high-stake environments where precise risk assessments are essential to minimizing harm. The following approaches are in no particular order. Choosing which method to use is based on the data availability, the safety problem, the familiarity of the analysis, and the desired output.

> *Probabilistic Risk Assessment* (PRA) is a quantitative method that estimates the likelihood and consequences of different hazards. By calculating the probability of various events, safety professionals can create a detailed picture of potential risks and rank them by severity. PRA is commonly used in industries such as nuclear energy and aerospace, where precise, data-driven risk assessments are critical for safety. This method's strength lies in its ability to provide a clear, numerical basis for decision-making, helping to effectively identify and mitigate high-probability hazards.

> *Cost–Benefit Analysis* (CBA) weighs the costs of implementing safety measures against the benefits, often measured in terms of reduced risks, fewer incidents, and increased compliance. CBA is a valuable tool for determining which safety interventions offer the highest return on investment, ensuring resources are allocated effectively. While this approach provides a financial lens on safety, its limitations include the potential underestimation of qualitative benefits, such as employee well-being or brand reputation. Despite these limitations, CBA remains a practical framework for evaluating safety initiatives in terms of both risk reduction and financial feasibility.

> *Failure Modes and Effects Analysis* (FMEA) systematically identifies possible failures in a process, assesses their causes, and estimates their potential impact. FMEA assigns a Risk Priority Number to each potential failure, enabling safety professionals to prioritize risks based on severity, likelihood, and detectability. FMEA is especially useful in manufacturing and engineering contexts where a detailed process analysis is necessary. Its structured approach helps identify weaknesses in operations, thereby supporting targeted interventions that address specific failure points.

QUALITATIVE APPROACHES

The following methodology is often overlooked by senior management due to the difficulty in measuring and the lack of reproducibility, but this approach has much value when effectively used, as this approach can often answer the "why" questions in safety. A qualitative approach to safety decision-making focuses on understanding the context and nuances of risk, which may not always be quantifiable. These methods are essential for situations where numerical data is limited, or the nature of the risk involves significant uncertainty.

Hazard and Operability Study (HAZOP) is a qualitative technique that systematically identifies potential hazards by analyzing deviations from normal operations.
By examining each process step and brainstorming possible deviations, safety
professionals can identify hidden risks and assess the impact of potential failures.
HAZOP is widely used in industries such as chemical manufacturing, where
understanding complex process interactions is essential for safety. This method
encourages thorough examination and fosters collaboration, as team members
contribute diverse perspectives on potential hazards.

Root Cause Analysis (RCA) seeks to identify the underlying causes of incidents
or near-misses to prevent recurrence. Safety professionals can investigate
past incidents by investigating systemic issues or flaws contributing to risks.
Techniques such as the "5 Whys" and fishbone diagrams help probe deeper
into incidents to find root causes rather than superficial explanations. RCA is
invaluable in risk management, as it shifts the focus from symptoms to underlying factors, encouraging long-term safety improvements rather than short-
term fixes.

SWOT Analysis (Strengths, Weaknesses, Opportunities, and Threats) provides a
holistic view of internal and external safety factors. Although more commonly
associated with strategic planning, SWOT can help safety professionals understand their organization's capacity to manage risk effectively. For example,
strengths might include a strong safety culture, while weaknesses could be
outdated equipment. Opportunities could involve adopting new safety technologies, and threats might include emerging industry risks. This framework enables
a broad perspective on operational and contextual factors influencing safety.

HYBRID APPROACHES: COMBINING QUANTITATIVE AND QUALITATIVE INSIGHTS

The hybrid or mixed method approaches in safety decision-making integrate quantitative data with qualitative insights to broadly understand risks. By blending these
methods, safety professionals can achieve a more comprehensive assessment of
hazards, combining numerical precision with context-based judgment.

Risk Matrix combines quantitative severity and probability ratings with qualitative
descriptions of risk categories, allowing for a straightforward visual representation of priority risks. Risks are placed on the matrix based on their likelihood
and impact, helping professionals quickly identify which risks need immediate
attention. The simplicity of the risk matrix makes it a widely used tool in many
industries, as it allows for intuitive prioritization and is easy to communicate to
stakeholders.

BowTie Analysis is a hybrid technique that maps out the relationship between
hazards, potential causes, and preventive controls. The BowTie diagram visually
represents the pathways leading to a hazard (causes) on one side and the controls
and consequences on the other side. By showing proactive and reactive measures,
BowTie analysis provides a structured approach to understanding how hazards
could escalate and what controls are in place. This tool is especially effective in

high-risk sectors like oil and gas, where both preventative and mitigative actions must be robustly managed.

Fault Tree Analysis (FTA) is a logical, structured approach that identifies potential causes of a specific failure event. Using a diagram resembling a tree, FTA maps out the contributing factors leading to a failure, providing both qualitative insight and quantitative probabilities. FTA is particularly valuable when investigating complex system failures, as it allows professionals to isolate root causes, understand the interdependencies among factors, and prioritize risk reduction measures based on likely impact.

SELECTING THE RIGHT FRAMEWORK

Choosing the right framework depends on the specific context, available data, and desired outcomes. Quantitative approaches are beneficial when dealing with high-volume data and when precision is critical, such as in regulatory compliance or financial risk analysis. Qualitative methods, however, are valuable for situations that require human insight, such as understanding process nuances or identifying potential hazards in new or complex environments. Hybrid approaches provide a balanced solution, integrating the strengths of both quantitative and qualitative methods.

ADAPTING FRAMEWORKS FOR EVOLVING SAFETY CHALLENGES

The rise of new technologies, regulatory changes, and evolving industry practices mean that frameworks for safety decision-making must be adaptable. Professionals should be prepared to refine or combine existing frameworks as needed, incorporating real-time data and insights from advanced tools such as predictive analytics and AI. For example, imagine how risk heat maps could be enhanced by incorporating live data from monitoring systems, enabling more dynamic assessments of evolving risks. Similarly, ML insights can enrich FMEA and HAZOP analyses that reveal patterns humans may overlook.

Ultimately, these frameworks foster a proactive risk management culture, enabling safety professionals to make well-formed decisions in complex, dynamic risk environments. With these frameworks as a foundation, safety professionals can effectively address immediate safety challenges and adapt to future technological advancements and risk management practices.

ASSESSING AND PRIORITIZING RISK IN SAFETY DECISION-MAKING

For safety professionals, accurately assessing and prioritizing risks is fundamental to managing safety effectively. With various hazards and operational uncertainties, professionals must evaluate which risks require immediate action and which can be managed over time. Prioritization maximizes resource efficiency and promptly addresses the most pressing safety issues. This section looks at essential steps in risk assessments, tools for evaluating risk severity, and strategies for prioritizing risks based on their potential impact.

LOGICAL STEPS IN THE RISK ASSESSMENT PROCESS

Risk assessment involves structured steps that help professionals understand, evaluate, and quantify risks. This systematic approach allows for consistent, comprehensive analysis and provides a foundation for prioritizing interventions. Though there are many ways to conduct a risk assessment, there is a general acceptance of the process:

1. Identify Hazards: The first step in any risk assessment is hazard identification. Safety professionals often conduct thorough investigations to locate potential sources of harm within their environment, processes, or systems. Hazards can stem from different factors, including equipment malfunction, human error, environmental conditions, or external threats like natural disasters. Effective hazard identification requires not only observational skills but also familiarity with the processes, equipment, and conditions present within the organization. Techniques such as process mapping, checklists, and historical incident reviews can support this step.

2. Determine Risk Consequences and Likelihood: Once hazards are identified, the next step is to assess each hazard's potential consequences and likelihood. Consequence assessment evaluates the possible outcomes if a risk materializes, such as injuries, fatalities, financial losses, or business interruption outcomes. Likelihood assessment, meanwhile, involves estimating the probability of the hazard occurring within a specific period. Remember, any event can happen with enough time, so specifying the period for the evaluation is essential. Probability matrices, fault tree analysis, and historical data analysis help quantify these aspects.

3. Evaluate Risk: With an understanding of consequences and likelihood, safety professionals can evaluate each risk's overall significance. This evaluation typically uses a combination of qualitative judgments and quantitative measurements to establish the relative importance of each risk. Evaluation frameworks, such as a risk matrix or scoring system, help classify risks as low, medium, or high priority based on their assessed impact and probability.

4. Develop Control Measures: After evaluating risks, professionals establish control measures to mitigate or manage them. Control measures fall into categories based on the "hierarchy of controls," which ranks them by effectiveness: elimination, substitution, engineering controls, administrative controls, and personal protective equipment (Manuele, 2005; Morris & Cannady, 2019). Higher-order controls, like elimination and substitution, are most effective but may not always be feasible. Safety professionals must balance effectiveness, feasibility, and cost when selecting appropriate control measures.

RISK PRIORITIZATION TECHNIQUES

In the previous section, we discussed risk assessments that help identify risks, so now we move to risk prioritization techniques. With a thorough evaluation of potential risks, safety professionals must prioritize which risks to address first. Prioritization is critical to effective resource allocation and ensures that high-risk areas are managed promptly. Here are some, but not all, standard techniques for prioritizing risk.

ALARP Principle (As Low As Reasonably Practicable): The ALARP principle is based on the idea that risks should be reduced to the lowest level reasonably achievable, balancing risk reduction with economic feasibility. This approach benefits high-risk industries, such as energy and transportation, where some hazards cannot be eliminated without eliminating the product. By focusing on what is "reasonably practicable," safety professionals can prioritize cost-effective, impactful measures that significantly reduce risk without excessive resource expenditure.

Cost–Benefit Analysis: CBA is another effective prioritization tool that weighs the expected cost of implementing a safety measure against the anticipated benefit in terms of risk reduction. This approach can help safety professionals allocate budgets efficiently, focusing on interventions with the greatest return on investment. While CBA is often quantitative, it should consider qualitative benefits, such as improved employee morale and strengthened regulatory compliance.

Criticality Analysis: Criticality analysis ranks risks based on their potential impact on core operations or objectives. This technique is especially relevant in industries where certain functions are essential to safety or productivity, such as emergency response or utilities. By identifying risks that could have the most severe consequences for these critical functions, professionals can ensure that safety measures focus on preserving operational continuity and resilience.

Multi-Criteria Decision Analysis (MCDA): MCDA is a decision-making framework that considers multiple factors when prioritizing risks, such as likelihood, severity, cost, feasibility, and stakeholder impact. By assigning weights to each criterion, safety professionals can score and rank risks objectively. MCDA's comprehensive, flexible approach allows for customization based on an organization's unique risk landscape and priorities, making it a valuable tool for balancing competing priorities.

Pareto Analysis (80/20 Rule): The Pareto principle, or 80/20 rule, suggests that 80% of the impact comes from 20% of the causes. In risk management, this means identifying the small subset of risks that contribute most significantly to potential harm. Pareto analysis helps professionals focus on the most impactful risks, enabling efficient use of resources by addressing issues with the most significant potential to improve safety outcomes.

These are some of the more common risk prioritization techniques, and each comes with strengths and limitations. It's encouraged that you read into these processes and develop a straightforward methodological approach when you engage in this risk assessment and prioritization process.

BALANCING IMMEDIATE AND LONG-TERM RISK PRIORITIES

In prioritizing risks, safety professionals must balance immediate threats against long-term issues. Immediate risks, which threaten safety or regulatory compliance, often demand urgent action, while long-term risks may be addressed during strategic planning. This balance requires a nuanced understanding of each risk's potential impact on current operations and future resilience. In industries where technological advancements and regulatory landscapes change frequently, professionals must periodically reassess and adjust priorities to ensure alignment with evolving conditions. Continuous monitoring and updating of risk assessments ensure that prioritization remains relevant and responsive to new data, incident learnings, and emerging threats.

IMPLEMENTING RISK-INFORMED SAFETY DECISIONS: STRATEGIES AND BEST PRACTICES

Once risks have been assessed and prioritized, the challenge lies in translating this understanding into actionable safety measures. Effective implementation of risk-informed decisions requires aligning safety practices with organizational goals, fostering a proactive safety culture, and monitoring outcomes for continuous improvement. This section explores three strategies for embedding risk-informed decisions within safety programs and discusses best practices to ensure they achieve their intended impact.

STRATEGY 1: ALIGNING SAFETY INITIATIVES WITH ORGANIZATIONAL GOALS

A foundational strategy for effective risk-informed decision-making is to align safety initiatives with broader organizational objectives. Safety decisions should support operational goals, regulatory compliance, and corporate values. This alignment creates a cohesive approach that benefits safety programs from organizational resources and attention.

Safety decisions are more likely to be adopted and practiced when integrated with operational performance objectives. Safety professionals can enhance performance while reducing risks by aligning safety measures with productivity, quality, and cost-efficiency goals. For instance, safety training focusing on operational efficiency (such as safe handling practices in manufacturing) reinforces the dual objectives of safety and productivity.

Incorporating safety considerations in strategic planning ensures that safety priorities are communicated and reinforced at all levels of the organization. Management must define safety goals as part of the strategic vision and integrate them into performance metrics, budgeting, and project planning. An organization with clear safety goals is better positioned to implement risk-informed decisions that align with its mission, vision, and values.

Allocating resources to high-risk areas identified in the risk prioritization process is crucial for effective implementation. Resources, including time, budget, and personnel, should be focused on the most critical safety interventions. Resource allocation should also be flexible to accommodate evolving risks and changing priorities, ensuring continuous support for essential safety measures.

STRATEGY 2: BUILDING A PROACTIVE SAFETY CULTURE

A proactive safety culture is essential for the successful implementation of risk-informed decisions. This culture is one in which employees at all levels understand, support, and engage in safety initiatives, creating an environment where safety is prioritized daily.

Education and training are central to building a safety culture that actively supports risk-informed decisions. Programs should be tailored to the organization's specific safety needs, focusing on general safety principles and specialized training related to prioritized risks. For example, training on hazard recognition and emergency response procedures equips employees to participate actively in risk reduction efforts. Education and training these days are often confined to virtual platforms. Encourage real-life training for critical practices, as the collaborative, hands-on, and team-focused approaches promote positive aspects of safety culture.

Actively involving employees in safety initiatives creates a sense of ownership and responsibility. Engagement can be achieved by including frontline workers in safety meetings, soliciting their input on risk assessments, and encouraging them to report near-misses and unsafe conditions. This process will also ensure diversity of thought and experience on the safety team. Remember, there is no reason to perform safety analysis alone, as this will create inherent biases. Engagement improves risk data quality and enhances employee commitment to implementing safety decisions.

Empowering employees to make real-time safety decisions reinforces a proactive safety culture. This empowerment involves providing workers with the authority, training, and resources to address immediate safety concerns. Employees who feel empowered to act on potential risks become active contributors to the organization's safety goals, reducing incidents through rapid, decentralized decision-making. Empowerment often promotes fresh ideas and creative problem-solving. The key to making this process work is to encourage open ideas without fearing reprisal if an idea does not work out.

STRATEGY 3: DEVELOPING A STRUCTURED IMPLEMENTATION PROCESS

A structured process for implementing risk-informed decisions is essential to translating risk insights into tangible safety improvements. By following a systematic approach, organizations can ensure that safety measures are executed effectively and consistently.

Safety protocols provide clear guidelines for implementing risk-informed decisions. These protocols should outline specific actions, responsibilities, and timelines for addressing each identified risk. Clear, well-documented protocols facilitate compliance and serve as a reference during training and daily operations. In high-risk environments, protocols may include detailed emergency response plans, hazard mitigation steps, and contingency measures. Protocols should be reviewed often to ensure they remain relevant to the tasks at hand, and each time a change occurs, be sure to document the new changes and preserve a record for future use.

Technology, including safety management systems (SMS), wearable devices, and data analytics platforms, supports the effective implementation of risk-informed decisions by tracking progress, identifying issues, and providing real-time insights. SMS software, for example, can manage safety documentation, training records, incident reports, and audit results, ensuring that safety data is accessible and actionable. Wearable devices and Internet of Things (IoT) sensors provide real-time data on environmental hazards, enabling timely responses to evolving risks. With the advancement of Bluetooth technology, many organizations are looking toward IoTs to promote operational efficiency. Therefore, safety managers should stay up to date with the new technologies and engage with the information technology teams to capture data opportunities.

Key Risk Indicators (KRIs) and metrics allow organizations to track the effectiveness of their risk-informed decisions. Metrics should be tied to specific safety objectives, such as reduced incident rates, compliance levels, or hazard response times. Monitoring KRIs helps safety professionals assess the impact of implemented measures and identify areas that may require adjustments. Regularly reviewing these metrics promotes a continuous improvement mindset and keeps safety performance on track with organizational goals.

BEST PRACTICES FOR EFFECTIVE IMPLEMENTATION

Pilot programs allow organizations to test new safety interventions on a small scale before full implementation. Pilots provide valuable insights into the feasibility and impact of specific measures, helping to identify potential issues and refine approaches. By piloting changes in selected departments or sites, organizations can assess the practical challenges of implementation and adjust protocols as needed for a smoother rollout. Be sure to document the various stages of pilot rollouts, as this information will become valuable in future segment implementation.

Effective safety implementation often requires collaboration across departments, including operations, human resources, and facilities management. Cross-functional collaboration ensures that safety decisions consider diverse perspectives, leading to more comprehensive safety measures. For example, collaborating with facilities management can support the effective rollout of engineering controls, while human resources can support safety training and compliance tracking. Building a team of safety champions will facilitate earlier and more effective adoption within the workforce.

Implementation is not a one-time effort but a continuous process that requires regular feedback, adaptation, and improvement. Establishing feedback loops allows employees to share insights on the effectiveness of safety measures and suggest adjustments based on their experience. Safety professionals can use incident data, audits, and employee feedback to refine protocols and ensure that safety practices remain relevant to evolving risks. Incorporating multiple communication channels during implementation will allow more data to be collected and, hopefully, more robust decision-making for optimal outcomes.

Leadership commitment is a crucial driver of successful implementation. Leaders must visibly support safety initiatives, communicate their importance, and provide the necessary resources for effective execution. Regular safety updates, transparent communication about risk decisions, and recognition of safety achievements all reinforce the organization's commitment to safety. When employees see that leaders prioritize safety, they are more likely to adopt and uphold safety practices. Any safety initiative without safety champions will likely fail due to the lack of management commitment. Make sure to engage management early and often to promote the right outcomes.

Regular audits ensure that safety decisions are consistently implemented and protocols remain effective. Audits can identify deviations from established procedures, assess compliance with regulations, and uncover areas for improvement. Safety professionals can maintain high safety performance standards and reinforce organizational accountability by conducting scheduled and surprise audits. Audits should not be used as a punishment but as a barometer of the safety culture. If the safety culture is excellent, the audit will likely be welcomed by employees, but if there is a poor safety culture, then the audit will likely be seen as a burden or hindrance. Understanding this culture is just as important as the results of the audit.

The dynamic nature of risks requires that organizations regularly reassess and update their safety measures. New information from incident reports, technological advancements, and regulatory changes should inform ongoing adaptations of safety practices. For example, emerging risks from automation or new machinery may require updating safety protocols, while predictive analytics can provide early warnings about potential hazards. Organizations can maintain a resilient and adaptive approach to safety by staying responsive to change.

MONITORING AND CONTINUOUS IMPROVEMENT OF SAFETY DECISIONS

Effective implementation involves continuous monitoring and improvement to ensure safety measures achieve their intended outcomes. This approach emphasizes the ongoing refinement of safety practices and adapts to changing circumstances, enabling organizations to respond proactively to new risks.

Real-time monitoring, supported by data analytics, allows safety professionals to track safety performance metrics and incident data as they occur. Advanced analytics can reveal patterns and trends, supporting proactive adjustments to safety measures. This data-driven approach provides early warning signals, allowing for timely interventions that can prevent incidents before they escalate. Furthermore, real-time monitoring will allow the operations team to adapt to changing conditions versus waiting for results to occur, which helps promote a resilient organization (Haynes & McAleavy, 2021).

Post-implementation reviews provide insights into the effectiveness of safety measures, revealing successes and areas for improvement. These reviews should assess whether the implemented measures achieved their intended impact, identify challenges encountered during implementation, and highlight lessons learned. Regular post-implementation evaluations create a feedback loop that promotes continuous improvement and enables future risk-informed decisions to be implemented more effectively.

Military organizations often use these reviews after operations, and each team member is expected to contribute. During my career, I would use the back of a truck bed to conduct these reviews, as they don't need to be highly formal with written reports. They need to be integrated into the standard practices of your teams to promote real change.

A learning culture supports continuous improvement by encouraging employees to learn from both successes and failures. Safety professionals can refine safety measures and share lessons across teams by analyzing incidents, near-misses, and feedback. A learning culture also fosters openness, where employees feel comfortable reporting safety concerns without fear of repercussions, ultimately enhancing the organization's ability to implement effective, risk-informed decisions. A critical component of a learning culture is creating learning knowledge banks. These knowledge banks help ensure that information is recorded, maintained, and available for future employees.

EVALUATING OUTCOMES AND PROMOTING CONTINUOUS IMPROVEMENT

Once risk-informed safety decisions are implemented, it is essential to evaluate their effectiveness and make continuous improvements to adapt to evolving risks. Evaluation and improvement processes help ensure that safety measures achieve their intended outcomes, identify areas for refinement, and foster a resilient, proactive approach to risk management. This section discusses methodologies for outcome evaluation, the importance of performance metrics, and strategies for embedding a culture of continuous improvement within safety programs.

IMPORTANCE OF EVALUATING OUTCOMES IN RISK DECISION-MAKING

Evaluating the outcomes of risk-informed decisions allows safety professionals to verify that implemented safety measures effectively mitigate identified risks. It also highlights any gaps or unforeseen issues that may require further action. Outcome evaluation is a critical step in closing the loop in the decision-making process, ensuring that safety practices are effective and adaptable to changing conditions.

Evaluation helps validate that risk mitigation strategies achieve their intended impact and capture unintended benefits, as this helps promote future safety initiatives. By comparing actual outcomes with expected results, safety professionals can confirm whether the implemented controls are sufficient. For example, if a new protocol is introduced to reduce slips and falls, monitoring incident rates before and after implementation can reveal its effectiveness. Validation ensures that limited resources are allocated to effective measures, reducing wasted effort on interventions that do not provide meaningful safety improvements.

Evaluation can also reveal gaps in safety measures or unintended consequences that may have emerged post-implementation. For instance, new safety equipment might inadvertently reduce productivity or create ergonomic challenges, leading to different types of risk. Identifying such issues allows for adjustments that improve safety without introducing new hazards. Regular evaluation uncovers these gaps, helping professionals refine their strategies to address organizational needs better.

Outcome evaluation generates data supporting future risk-informed decisions and helps professionals better understand what works in their specific context. Organizations can establish evidence-based safety practices by tracking trends and analyzing data from incidents, near-misses, and safety audits. This data-driven approach provides objective insights, reducing reliance on anecdotal evidence or sub-jective judgments and promoting consistency in risk management decisions. Data collection also fosters the application and use of future technologies in your organization, and management is likely to see your team as an asset if you are correctly gathering and storing the information.

PERFORMANCE METRICS AND KEY INDICATORS IN SAFETY

To effectively evaluate safety outcomes, safety professionals must rely on per-formance metrics and key indicators that provide quantifiable data on the success of their interventions. These metrics allow for objective assessment, helping deter-mine whether safety measures meet predefined targets and goals. Some of these metrics are:

Lagging and Leading Indicators: Lagging indicators are retrospective metrics that measure past performance, such as incident rates, injury severity, and lost workdays. While helpful in assessing outcomes, lagging indicators do not pro-vide early warning signals for potential issues. Conversely, leading indicators measure proactive efforts, such as safety training completion rates, near-miss reporting, and audit scores. Leading indicators allow organizations to monitor preventive activities as early predictors of potential risks before they result in incidents.

Establishing KRIs Aligned with Safety Goals: KRIs should be directly tied to an organization's safety goals and risk priorities. For example, if reducing machinery-related injuries is a priority, a relevant KRI might track the frequency of equipment maintenance checks. By aligning KRIs with specific objectives, organizations can focus on metrics that reflect actual safety performance, cre-ating a clear link between risk decisions and operational outcomes.

Benchmarking against Industry Standards: Comparing performance metrics with industry benchmarks provides additional context for evaluating outcomes. Industry standards, such as incident rates in similar organizations or compliance metrics from regulatory bodies, help safety professionals assess their perform-ance relative to peers. Benchmarking highlights areas where the organization excels or lags, offering insights into best practices and improvement opportun-ities. Additionally, industry benchmarks can serve as reference points for setting realistic targets and goals within the organization. Industry conferences and journals are great places to hear about these standards.

Real-Time Monitoring and Data Analytics: Real-time monitoring and analytics tools provide dynamic insights into safety performance. Technologies like IoT devices and wearable sensors capture data on environmental conditions, worker behavior, and equipment performance. This real-time data allows for immediate responses to emerging risks and enables predictive analytics to forecast poten-tial issues based on historical trends. Organizations can continuously monitor key metrics to detect deviations from safety standards and intervene before risks escalate.

These are some of the metrics that organizations may use to measure performance. Consider asking teams around your company how they capture performance. You might find another avenue to improve safety performance and data collection.

FOSTERING A CULTURE OF CONTINUOUS IMPROVEMENT

Creating a culture of continuous improvement requires ongoing commitment from both leadership and employees. Safety professionals should emphasize the value of improvement over punitive responses to safety incidents, encouraging openness and collaboration. Employees will likely follow safety rules to preserve their jobs and avoid harm, but mandating safety does not promote organizational resilience. Instead, safety teams could promote continuous improvement through leadership, recognition, and familiarity.

Leaders play a vital role in promoting continuous improvement by setting an example and prioritizing safety. Transparent communication from leadership about safety improvements and recognition of successful initiatives reinforce a commitment to safety at all levels. When leaders visibly support continuous improvement, employees are more likely to embrace it as part of the organizational culture. Leaders attract followers, who will likely follow if the leader promotes safety.

Recognizing employees who actively contribute to safety improvements encourages ongoing engagement. By acknowledging achievements, such as reduced incident rates or innovative safety suggestions, organizations motivate employees to participate in continuous improvement. Rewards can include formal recognition, incentives, or public acknowledgment, creating a positive reinforcement cycle that promotes ongoing commitment to safety enhancement. Consider working with a marketing or human resource employee to find other creative ways to encourage employee recognition—they are usually creative people who enjoy fostering healthy employee cultures.

Finally, continuous improvement should be embedded into daily operations, not treated as a separate initiative. Integrating safety discussions into team meetings, incorporating improvement goals into performance reviews, and regularly updating safety training based on recent evaluations all contribute to a culture of ongoing enhancement. When improvement processes are a part of routine operations, they become second nature to employees, supporting a sustainable approach to risk-informed decision-making. Safety is not something we do, but it is how we behave. Promoting continuous improvement is no different from an always-learning mindset.

FUTURE DIRECTIONS: INTEGRATING INNOVATION AND FOSTERING RESILIENCE

As organizations increasingly rely on advanced technologies and data-driven approaches to enhance safety, the future of risk decision-making will transform dynamically. Emerging trends such as automation, AI, predictive analytics, and integrated risk management systems will shape how safety professionals approach risk-informed decisions. This segment surveys the potential impacts of these innovations, offering a forward-looking perspective on the evolving landscape of safety management and its implications for risk decision-making.

THE ROLE OF AUTOMATION AND AI IN SAFETY MANAGEMENT

Automation and AI are two of the most transformative trends in safety management, offering opportunities to improve the speed, accuracy, and predictive capability of risk assessments and safety interventions. To cover all the potential applications and risks for safety professionals would take an entire book of its own. Instead, I will provide a brief snapshot of the major technologies, applications, and perspectives for the safety professional.

> *AI for Predictive Risk Assessment*: AI algorithms can process vast amounts of data to identify patterns, trends, and early warning signs of potential hazards. Predictive analytics powered by ML enables safety professionals to anticipate risks before they lead to incidents. For example, AI can predict equipment failures in industries like manufacturing and construction by analyzing real-time sensor data, allowing maintenance to be scheduled proactively. This predictive approach reduces downtime, improves safety, and enables more targeted risk management.
>
> *Automation in Hazard Monitoring and Incident Response*: Automation allows organizations to monitor hazards and respond to incidents efficiently. Automated safety monitoring systems, such as IoT-enabled sensors, can detect hazardous conditions (like toxic gas levels, high temperatures, or unsafe machine conditions) and automatically trigger alarms or shut down equipment. This capability minimizes human intervention and ensures timely responses, reducing the likelihood of injuries and operational disruptions.
>
> *AI-Driven Decision Support for Safety Professionals*: AI-driven decision support systems provide safety professionals with actionable insights based on real-time data, historical trends, and risk assessments. These systems can generate recommendations for risk controls, offer scenarios for different mitigation strategies, and simulate the potential impact of safety interventions. By integrating AI-driven support, safety professionals can make faster, more informed decisions grounded in data, enhancing the effectiveness of safety measures.

As this field is rapidly evolving, a common place to hear about these innovations and tools is through conferences and industry publications. Most people don't enjoy advertisements, but I would encourage you to stop occasionally and ask how this might be useful in my organization. These "How" questions can often start deeper conversations on improvements.

ENHANCING SAFETY THROUGH ADVANCED DATA ANALYTICS

Advanced data analytics, including big data and ML, empower organizations to adopt a data-centric approach to safety. By collecting, processing, and analyzing extensive datasets, organizations can gain deeper insights into risk factors, enabling more precise and proactive safety measures. The process is generally similar in most organizations: what data do we have, utilize an analytical tool and approach, and study the results for implementation.

Big data involves gathering large volumes of structured and unstructured data from various sources, such as equipment sensors, employee feedback, and incident reports. Analyzing this data provides insights into the root causes of incidents, high-risk activities, and safety trends. For example, analyzing data on near-misses across multiple facilities can reveal common factors contributing to these incidents, leading to targeted interventions. Big data enables a comprehensive view of safety performance, supporting evidence-based decision-making and continuous improvement.

ML algorithms can identify hidden patterns within data that traditional analysis methods might miss. By continuously learning from new data, these algorithms become more accurate over time, enhancing their ability to predict risks. ML can also help organizations personalize safety measures based on individual risk profiles, such as workers' skill levels or exposure to specific hazards. This personalized approach reduces risks and creates a safer work environment tailored to the unique needs of each employee.

It's important to know that a limited number of rules often define how the machine works, and that ML is an iterative process, meaning that each time you submit data, it is used for future analysis. This constant data recycling may cause an algorithm drift to appear, and special techniques must be implemented to ensure the machine operates as intended. ML must include a human-in-the-middle when it comes to safety. This tool dramatically impacts all business areas but ensures that ML is a supportive tool, not a replacement tool for a great risk analyst.

Integrating analytics into risk decision-making enables safety professionals to shift from reactive to proactive risk management. Real-time analytics dashboards provide continuous visibility into safety metrics, alerting professionals to emerging risks. Additionally, analytics-based decision-making frameworks allow organizations to assess the likely outcomes of various interventions, making selecting the most effective risk control measures easier.

Fostering Organizational Resilience and Adaptability

As the pace of change accelerates, resilience and adaptability will be crucial qualities for organizations seeking to maintain high safety standards in uncertain environments. Future risk decision-making must consider how organizations adapt to new risks and quickly recover from disruptions. The new wave of safety management will appear vastly different, but the process will generally be the same. Promoting resilience starts with planning. Resilience is reinforced by fostering a learning culture, and the approach will primarily be human-centric.

Scenario planning is a tool for preparing organizations to respond effectively to a range of possible futures. Many organizations are already planning scenarios at the strategic level. By simulating different risk scenarios at the operational level—such as natural disasters, equipment malfunctions, or cyberattacks—organizations can assess the resilience of their safety programs. Scenario planning encourages flexibility, enabling organizations to respond proactively to unexpected risks. This approach enhances preparedness and builds a culture of adaptability among employees, essential for managing complex safety challenges.

The future of risk decision-making relies on organizations fostering a culture of continuous learning and improvement – have you picked up on the common theme of this chapter? This culture values feedback, incident analysis, and employee input as sources of learning and improvement. Safety programs that promote continuous learning are better equipped to adapt to evolving risks, ensuring that safety measures remain relevant. Continuous learning also helps organizations identify and integrate emerging best practices, keeping their safety programs aligned with the latest industry standards.

A human-centered approach to designing safety systems considers employees' needs, limitations, and preferences. For example, user-friendly interfaces for safety monitoring tools, ergonomic equipment design, and intuitive training materials enhance the usability and acceptance of safety measures. Organizations can prioritize human-centered design to improve employee engagement with safety programs, leading to more effective risk control measures and a more robust safety culture. Employees will likely be apprehensive about new technologies but engaging them early in the process may foster adoption and acceptance.

KEY TAKEAWAYS AND THE ROAD AHEAD

The evolution of risk decision-making in safety management reflects broader trends toward data-driven processes, automation, and organizational resilience. Safety professionals who embrace these innovations will be better positioned to address complex and dynamic risks in the workplace. However, as these technologies are integrated into safety programs, organizations must also remain mindful of ethical considerations, employee engagement, and the potential unintended consequences of overreliance on automation.

While technology offers significant advantages, human expertise and judgment remain irreplaceable components of effective safety management. Safety professionals must balance automated decision-support systems with their experience and knowledge, ensuring that technology can enhance—not replace—human insight. AI will never be able to replace a collaborative safety team, as the value of experience and human insights is not reproducible by a machine. ML is excellent for summarization and data analysis, but the "so what" question can only be answered by us—whether we are management, the safety professional, the employee, or any other stakeholder.

By integrating these principles, strategies, and forward-looking insights, safety professionals can approach risk decision-making with a robust toolkit for managing risks effectively in today's dynamic environment. Embracing both technological advancements and human-centered values, safety professionals can foster safer workplaces, support organizational resilience, and build a proactive safety culture that prepares their organizations for future challenges.

REFERENCES

Ben-Haim, Y. (2012). Doing our best: Optimization and the management of risk. *Risk Analysis: An International Journal, 32*(8), 1326–1332.

Bilalić, M., Langner, R., Erb, M., & Grodd, W. (2010). Mechanisms and neural basis of object and pattern recognition: A study with chess experts. *Journal of Experimental Psychology: General, 139*(4), 728.

Cassell, C. A., Dearden, S. M., Rosser, D. M., & Shipman, J. E. (2022). Confirmation bias and auditor risk assessments: Archival evidence. *Auditing: A Journal of Practice & Theory, 41*(3), 67–93.

Folkes, V. S. (1988). The availability heuristic and perceived risk. *Journal of Consumer Research, 15*(1), 13–23.

Haynes, S., & McAleavy, T. (2021). Integrating local personnel response and recovery capacity: A conceptual model for small to medium enterprise hazard risk analysis. *Journal of Business Continuity & Emergency Planning, 15*(1), 1–18.

Kalantari, B. (2010). Herbert A. Simon on making decisions: Enduring insights and bounded rationality. *Journal of Management History, 16*(4), 509–520.

Lieder, F., Griffiths, T. L., M. Huys, Q. J., & Goodman, N. D. (2018). The anchoring bias reflects the rational use of cognitive resources. *Psychonomic Bulletin & Review, 25,* 322–349.

Manuele, F. A. (2005). Risk assessment & hierarchies of control. *Professional Safety, 50*(5), 33–39.

Morris, G. A., & Cannady, R. (2019). Proper use of the hierarchy of controls. *Professional Safety, 64*(08), 37–40.

Samuelson, W., & Zeckhauser, R. (1988). Status quo bias in decision making. *Journal of Risk and Uncertainty, 1,* 7–59.

Schwartz, B., Ben-Haim, Y., & Dacso, C. (2011). What makes a good decision? Robust satisficing as a normative standard of rational decision making. *Journal for the Theory of Social Behaviour, 41*(2), 209–227.

Simon, H. A. (1955). A behavioral model of rational choice. *The Quarterly Journal of Economics,* 99–118.

Simon, H. A. (1979). Rational decision-making in business organizations. *The American Economic Review, 69*(4), 493–513.

Whittaker, A. (2018). How do child-protection practitioners make decisions in real-life situations? Lessons from the psychology of decision making. *The British Journal of Social Work, 48*(7), 1967–1984.

6 Addressing Psychosocial Hazards Can Lead to Equitable Health and Safety

I. David Daniels

UNEQUITABLE OUTCOMES

Plenty of data suggests that many safety management systems do not provide equitable outcomes for all members of an organization. Interestingly, the discrepancies in these systems are often overlooked or discounted for various reasons. A hazard left unmitigated in an organization will increase in its frequency of occurrence, severity of harm, and duration of prevalence in the culture. If it is not safe for everyone, it is not safe for anyone. Hazards do not discriminate regarding their effect when a member is vulnerable. When there is inequitable treatment within the organization, it results in its various members' inequitable exposure to recognized hazards.

According to available data, significant demographic disparities exist in workplace injury and fatality rates. Based on race and ethnicity, black and Hispanic members experience considerably higher fatality rates compared to white members; specifically, black members often have the highest fatality rate among minority groups, followed by Hispanic members, while Asian members generally have the lowest rate (Jetha et al., 2023). Black and Hispanic members have a higher prevalence of injuries and disabilities than non-Hispanic white members. Black or African Americans have the highest rates of members' compensation claims across all industries and occupations. In unadjusted models, white members had lower odds of disability from a workplace injury than black or Asian members. Minority racial and ethnic groups are less likely to report injuries compared to non-Hispanic white members. While rates of injury vary across different population groups, overall, injuries are more common in men than in women. This higher rate of injury can be explained, at least partly, by the fact that men are more likely to engage in risk-taking behaviors and activities. We have all seen Facebook memes of the significant risk tolerance differences between the two genders.

Data points regarding the fatalities and injuries based on race, ethnicity, and gender indicate that our safety systems in the United States have output, outcome, and impact issues. However, there are also input issues with the systems that produce

DOI: 10.1201/9781003583103-7

these results. Credited with first introducing the concept of human organizational performance in the mid-20th century, Dr. W. Edwards Deming said, "Every system is perfectly designed to get the result that it does." Deming estimated that 94 % of the problems in a business are created by systems that aren't functioning as needed rather than individual actions. Interestingly, while people might apply a systems thinking, total quality approach to their business efforts, based on Deming's work, many don't use the same approach regarding their safety management systems. Too often, failures in safety are attributed to the human beings who are injured or killed as opposed to the systems that surround them.

HOMOGENEOUS FOUNDATIONS

A typical quote that captures the idea of seeing the world through our perspective is, "We see the world not as it is, but as we are," often attributed to Anaïs Nin and Stephen Covey. This quote highlights that our own experiences and biases filter our perceptions. Within a group setting, differences in background, perspective, and experiences frequently generate a more comprehensive range of ideas and, thus, greater deliberation than in a homogeneous group. Given a brainstorming task, diverse groups produced more practical and feasible ideas than homogeneous groups. This occurrence is not only actual in business but, given the data, also true in health and safety.

When the Occupational Safety and Health Act was signed into law in 1970, it was difficult to argue that lawmakers who approved the legislation could have considered the safety of all members. The 91st Congress was nothing if not homogeneous. New York representative Shirley Chisolm was a new member of the 91st Congress, and a member of two minority groups in the Congress. She was one of only 11 women (ten in the House and one in the Senate, 2.05%) and ten black people (nine in the House and one in the Senate, 1.86%). In other words, the 91st Congress was 97.95% male and 98.14% white. This reality is not intended to suggest that these public servants lack an interest in occupational safety but to suggest that they lacked the background, perspective, and lived experiences of women and people of color. Since the passage of the OSH Act, there has only been one amendment in 1998, which expanded the coverage to include members of the U.S. Postal Service. The relative lack of change in the law suggests that the original intent of the OSH Act continues today.

Despite gender and racial discrepancies, overall, the OSH Act has undoubtedly contributed to the reduction in workplace injuries and fatalities. In 1970, with a workforce of about 83 million, an estimated 14,000 members were killed on the job, averaging about 38 deaths a day. In 2022, with a workforce of 169.8 million, there were 5,486 fatal workplace injuries or 15 deaths a day. In 2022, private industry members reported 2.8 million nonfatal workplace injuries and illnesses. The injury rate was 2.3 cases per 100 full-time equivalent members, the same as in 2021. Overall, it is not hyperbole to argue that the workplace is demonstrably safer than when the OSH Act was adopted. However, it is also true that women and people of color continue to lag behind male and white members.

While a homogeneous group implemented the OSH Act, the safety profession is demographically similar to the overall workforce in terms of race. According to the 2020 U.S. Census, 61.6% of the population of this country was white, 18.9%

Hispanic, and 12.4% black. In 2023, the U.S. labor force was 76.5% white, 12.8% black or African American, 18.8% Hispanic or Latino, and 6.9% Asian. The safety profession was 72.2% white, while black, Asian, and other races were slightly better represented than their overall percentage in the workforce. Women are significantly underrepresented in the safety profession based on the fact that women are 46.9% of the workforce but only 31.1% of the safety profession.

REDUCING SAFETY DISPARITIES

Reducing disparities in workplace injuries and fatalities among women and people of color is a critical issue that the safety profession must address. These disparities often stem from different communication styles, cultural differences, and varying access to resources and training. One of the primary steps in addressing these disparities is implementing inclusive safety policies. Safety policies should be designed to consider the specific needs of diverse groups.

This step includes developing culturally responsive training programs that are sensitive to different cultural backgrounds and available in multiple languages. Additionally, establishing anonymous reporting systems can encourage members to report safety concerns without fear of retaliation. Regularly reviewing and updating safety policies ensures they remain equitable and effective. These policies are most effective when developed with a clear connection to best practices instead of being created primarily for inclusion.

Ultimately, policies designed for the objective purpose of improving and maintaining the health and safety of everyone in the organization will have the primary effect of reducing injuries and fatalities, with the collateral impact of helping those who have been traditionally left out. A crucial element of a safety management system's effectiveness is its apparent connection to its performance management system. While safety is undoubtedly a priority, performance management is needed to connect what an organization says about safety and what it does.

Ideally, every organization would strive for parity between their employee racial makeup and that of the general population of the community that the organization draws from and intends to serve. The next step would be to ensure that all levels of the safety team, including leadership, management, and technical staff, reflect the organization's diversity. This will help create an environment where all members feel valued and heard. These efforts should include positions such as safety technicians, officers, managers, and directors.

Providing equitable access to safety equipment and training for all members is crucial. Mentorship programs can also support the career development of women and people of color, helping them advance within the organization and contribute to a safer workplace. Engaging the workforce in safety initiatives can lead to better outcomes. Establishing safety committees that include representatives from all groups of employees ensures that different perspectives are considered in safety planning and decision-making. Regularly soliciting feedback from members on safety practices and concerns can help identify areas for improvement.

Implementing incentive programs to encourage safe behavior can motivate members to prioritize safety. Inclusion involves collaboration with external organizations that can provide additional resources and support for reducing disparities in workplace safety. Partnering with advocacy groups that focus on the rights of women and people of color can offer valuable insights and resources. Participating in government and philanthropic programs to improve workplace safety can provide support and funding. Engaging with industry associations to share best practices and resources can help organizations stay informed and adopt effective safety strategies.

Accurate data collection and analysis are vital for identifying and addressing disparities. If safety professionals can filter the collected data on injuries and fatalities by gender, race, and ethnicity, analyzing this data can help identify trends and areas of concern, allowing organizations to set specific targets for reducing disparities and track progress over time. This data-driven approach ensures that interventions are targeted and effective. Leveraging technology can significantly improve workplace safety. Wearable devices can monitor members' health and safety in real time, providing valuable data to prevent injuries. Safety apps can offer easy access to safety information and allow for the quick reporting of hazards. Virtual reality training can provide immersive and compelling safety training experiences, helping members better understand and respond to potential risks. It's important that all data collection efforts are used transparently, and as non-punitively as possible.

PSYCHOSOCIAL HAZARDS

Promoting psychological safety by creating an environment where members feel safe to speak up about safety concerns is essential. However, psychological safety is not attainable and cannot be maintained without addressing psychosocial hazards in the organization. This issue is crucial for those whose lived experience includes exposure to psychosocial hazards, such as bullying, harassment, and violence in the workplace. Given the polarized society in which many of us find ourselves today, these psychosocial hazards can be based on any number of criteria, such as race, political views or affiliations, gender, religious beliefs, language barriers, and so on. Psychosocial hazards are psychosocial factors that are perceived or experienced as unfavorable and can cause damage. Damage from a psychosocial hazard is in the mind of the individual exposed, which in turn influences their behavior. Understanding the definition requires understanding each component of the definition.

The first is the concept of psychosocial factors, which are influences that affect a person psychologically or socially. There are multidimensional constructs encompassing several domains, such as mood status (anxiety, depression, distress, and positive affect), cognitive behavioral responses (satisfaction, self-efficacy, self-esteem, and locus of control), and social factors (socioeconomic status, education, employment, religion, ethnicity, family, physical attributes, locality, relationships with others, changes in personal roles, and status).

The second component is the perception or experience of the person in proximity to the factor. How do the people closest to whatever is happening feel about it? This component is potentially the most crucial of the overall definition. Our brains make sense of the world around us by creating stories about the environment, predicting how these stories will affect us, acting on these predictions, evaluating our stories' accuracy, and starting the process all over again. An identical set of data points can be perceived or experienced very differently by different people. This component is undoubtedly the most significant as it relates to diverse populations. Too often, safety-related ideas, concepts, and systems are designed homogeneously and applied hetero-geneously. Unfortunately, due to differences in the perception of the words or actions, those exposed will experience a feeling that may or may not have been intended.

The third component is that the perception or experience of the factor is, at least in the mind of the person, one of threat or danger. One of the stories that our brains tell us based on our perceptions and experiences is that something in the environment is likely to bring harm to us. This reaction is a part of human nature and is essential in keeping us safe. The most impactful perceptions tend to be connected to our experi-ence of pain, hurt, or trauma. Situations that resemble these prior feelings will impact how we react to what we hear, see, smell, taste, or touch in our environment. The brain receives information about a possible threat from the five senses. The amyg-dala, the brain's emotional center, processes information and determines whether it's a threat. If it is a threat, the amygdala activates the hypothalamus, which triggers the sympathetic nervous system to release stress hormones like adrenaline and cor-tisol. Stress hormones prepare the body to fight or flee by increasing heart rate, blood pressure, and energy levels.

Finally, the fourth component is that the threat or danger is of a level or magnitude that influences the person's behavior. While an assortment of threats always exists in the environment around us, our perception of the degree to which the potential harm is controlled factors into our reactions. An example is any building constructed of combustible material that has the potential to catch fire. However, this potential is mitigated by fire sprinklers that minimize the potential for a fire to get large enough to create a significant threat. When we perceive a potential threat to be uncontrolled and feel vulnerable to potential harm, we are more likely to act in a way that creates distance between us and the threat or minimizes the harm we will experience. Though our bodies may prepare us to take action, the decision-making process results from communication between the prefrontal cortex (working memory) and the hippo-campus (long-term memory).

If any component is missing, the factor is not a hazard, as it does have the poten-tial to lead to harm, which is an essential component of the definition of a hazard. This definition opens the potential for anything to be a psychosocial hazard. These components are necessary because a workplace factor without some form of poten-tial to harm is rarely considered a psychosocial hazard. One person may recognize the factor; another may not notice it. One person may see it as a threat; another may not. One person may see the threat as significant enough to warrant a change in their behavior; another sees it as barely a nuisance.

INCLUSIVE LEADERSHIP

Little builds trust in an organization more than consistent leadership, where everyone feels included in the organization's direction. Inclusive leadership is a transforma-tive approach emphasizing the importance of all voices in fostering a positive organ-izational culture. This leadership style is crucial for driving organizational cultural change, as it ensures that all members feel valued, respected, and empowered to contribute their unique perspectives and talents. Inclusive leadership revolves around creating an environment where everyone feels they belong. Leaders who practice inclusivity know their biases and take deliberate steps to mitigate them. They demonstrate humility, empathy, and a genuine commitment to understanding and valuing the experiences of others.

Research has identified several vital behaviors that inclusive leaders exhibit. These include visible commitment, humility, awareness of bias, curiosity about others, cultural intelligence, and effective collaboration. Inclusive leaders prioritize these values in their decision-making processes and communicate their importance to the organization. They recognize their limitations and are open to feedback, understanding that they do not have all the answers and are willing to learn from others. Inclusive leaders are aware of their biases and actively work to counteract them, seeking to create a fair and equitable environment for all members. They show a genuine interest in understanding the perspectives and experiences of others, which helps them build stronger, more inclusive teams. Additionally, they can understand and navigate different cultural contexts, a skill essential for leading teams effectively. Finally, they foster a collaborative environment where all team members feel valued and heard, leading to better decision-making and innovation.

Inclusive leadership has a profound impact on organizational culture. Inclusive leaders help break down barriers and create a more cohesive and productive workforce by fostering an environment of respect and belonging. Members who feel included are likelier to be engaged, motivated, and committed to their work. Moreover, inclusive leadership drives innovation and brings various perspectives and ideas, which can lead to more creative solutions and better problem-solving. Members who feel their unique contributions are valued are likelier to share their ideas and collaborate effectively.

Several strategies can be employed to cultivate inclusive leadership within an organization. Providing training on these topics can help leaders develop the skills and knowledge needed to lead inclusively, including training on unconscious bias, cultural competence, and effective communication. Establishing mentorship and sponsorship programs can help underrepresented members advance in their careers, with inclusive leaders playing a crucial role by providing guidance and support. Supporting employee resource groups (ERGs) can provide a platform for members to connect, share experiences, and advocate for change, with inclusive leaders championing these groups and ensuring they have the resources they need to thrive.

Inclusive leadership is not just a moral imperative but a strategic advantage. By embracing these principles and fostering an inclusive culture, organizations can unlock the full potential of their membership. Inclusive leaders are pivotal in driving this cultural change, creating environments where all members feel valued, respected, and empowered to contribute their best work. As organizations continue to navigate an increasingly globalized world, the importance of inclusive leadership will only continue to grow.

Organizations that are interested in addressing the existing disparities relative to total members' health will only be ultimately successful when everyone is considered. This assertion is not intended to suggest that any system will be perfect and will meet every need, but if the priority is on the safety and total health of the human beings that the organization is focused on collaborating with and providing goods and services to, not only will the organization achieve its safety goals, but it will also achieve its performance, production, and profit goals.

BIBLIOGRAPHY

Canadian Centre for Occupational Health and Safety. Hazards. Retrieved on May 9, 2024 from www.osha.gov/safety-management/hazard-Identification#:~:text=Health%20hazards%20include%20chemical%20hazards,%2C%20repetitive%20motions%2C%20vibration). www.ccohs.ca/topics/hazards

Cornell Law School. 18 U.S. Code § 1589 – Forced labor. Legal Information Institute. Retrieved on May 9, 2024 from www.law.cornell.edu/uscode/text/18/1589

Daniels, Ira David. The lived experience of black workers' exposure to psychosocial safety hazards in the American workplace. PhD diss., Capitol Technology University, 2022.

EcoOnline. Health & safety glossary. Retrieved on May 7, 2024 from www.ecoonline.com/glossary/hazard

HandWiki. "Hazard" Encyclopedia. Retrieved on May 8, 2024 from https://encyclopedia.pub/entry/31441

International Federation of the Red Cross. Hazard definitions. Retrieved on May 7, 2024from www.ifrc.org/document/hazard-definitions

Jetha, A. et al. The future of work in shaping the employment inclusion of young adults with disabilities: A qualitative study. *Equality, Diversity and Inclusion,* 42(9), 2023, 75–91.

Koop L.K., and Tadi P. Neuroanatomy, sensory nerves. [Updated 2023 Jul 24]. In: StatPearls [Internet]. Treasure Island (FL): StatPearls Publishing; 2024 Jan. Available from: www.ncbi.nlm.nih.gov/books/NBK539846/

Mishra, T. Hazard. Safeopedia. Retrieved on May 7, 2024 from www.safeopedia.com/definition/152/hazard

National Association of Safety Professionals. Types of hazards. Retrieved on May 9, 2024 from https://naspweb.com/blog/types-of-hazards/

Occupational Safety and Health Administration. Hazard Identification and Assessment. Retrieved on May 9, 2024, from www.osha.gov/safety-management/hazard-Identification

Safe Work Australia. Retrieved on May 9, 2024 from www.safeworkaustralia.gov.au/safety-topic/hazards

Sugg, A. K., and Carrick, D., Zagi Kozarov sued after her job gave her PTSD. Her case is groundbreaking for those who do stressful work. *ABC News.*

Suzuki, Si., and Takei, Y. Psychosocial factors and traumatic events. In: Gellman, M.D., and Turner, J.R. (eds.), *Encyclopedia of Behavioral Medicine.* Springer, New York, NY. https://doi.org/10.1007/978-1-4419-1005-9_1716, 2013.

Tamrin, Shamsul Bahri Mohd, and Ishkandar Md Yusoff. 1 Hazards in workplace. *Occupational Safety and Health in Commodity Agriculture: Case Studies from Malaysian Agricultural Perspective* (2014): 55.

Trachsel LA, Munakomi S, and Cascella M. Pain theory. [Updated 2023 Apr 17]. In: StatPearls [Internet]. Treasure Island (FL): StatPearls Publishing; 2024 Jan. Available from: www.ncbi.nlm.nih.gov/books/NBK545194/

United Nations Office for Disaster Risk Reduction. Sendai framework terminology on disaster risk reduction. Retrieved on May 7, 2024.

7 Safety and Emotions

Simon Goncharenko and
C. Thomas Goncharenko

INTRODUCTION

All the other chapters in this book are considering various aspects of physical safety, leadership, and soft skills that can contribute to it or detract from it, and psychosocial elements that have a bearing on the culture, the lagging indicators, and company success. In this chapter, we want to consider the emotional side of our workforce, the challenges in this space, and the tremendous impact that this invisible element can have in your overall organizational journey. In the first draft of this chapter, we called emotions an "invisible hazard" in the previous sentence. However, upon further consideration, we had to rephrase it to an "invisible element" because emotions can be either a powerful motivator or a significant hazard, depending on how we understand and utilize them, which, of course, is one more reason for the need of this chapter.

Perhaps Dale Carnegie's reminder is appropriate here. In his *How to Win Friends and Influence People*, which has become the bestseller of all time, Carnegie proposes that "When dealing with people, … we are not dealing with creatures of logic" but "with creatures of emotion." Carnegie's point is supported by science, namely that the human brain is hardwired to use emotion in the decision-making process. That is why individuals that have sustained injuries to the limbic system of their brains lose the ability to make decisions.

And we think on some subliminal level we have appreciated this point for quite some time, with the emotional presentations and keynote speeches of individuals like Brad Livingston, Charlie Morecraft, Lee Shelby, Brandon Schroeder, and others being so prevalent today. These and many other individuals like them either survived a terrible industrial accident or are surviving family members whose experiences, testimonies, and stories tug on the emotions of their audiences and become powerful motivators for safer performance. These speakers are successful because they understand that to move their audiences to action, they must first make an emotional connection with them. Every volume on sales training also makes this same point.

Yet, at the same time, we do not fully understand or appreciate the value of those same emotions or the impact that they make on our workplaces, nor do we know how to harness this powerful tool, in order to make positive changes in our organizational culture. Another name by which this topic is also known is emotional intelligence or

DOI: 10.1201/9781003583103-8

EQ. And, while a lot has been written on this topic, including Chapter 3 of this book, an average worker in our experience still has much to learn about this area, to grow in mastering their emotions, and to practice applying these learnings to the interactions with others. Most people still do not recognize that if they take time to think through their emotions, they can manage them and understand their effects on others (Tyson, *The Human*). And, if application of this knowledge on a personal level is still in the infantile stage, then the connection between emotions or emotional intelligence and safety is even more enigmatic. This is precisely the reason that we need to discuss emotions and safety or emotional safety.

There are other reasons too. The second reason why this discussion is necessary is because an average individual does not fully appreciate the role that emotions play in our daily decision-making. While we all earnestly desire to believe ourselves to be generally rational people, in reality emotions play a far greater role in our decisions than what we give them credit for. In fact, as stated above, multiple psychological studies suggest that without emotions we would never make any decisions and that many if not most of our decisions are formed by emotions and then supported by logic.

The third reason why it is important to discuss emotional safety is because emotional safety and psychological safety are often either confused with one another or thrown into the same bucket. The truth of the matter is, however, that, while emotional safety is related to psychological safety and while both are essential for creating a healthy workplace environment that supports employee growth and development, the two are not the same. And we need to stop the confusion.

The fourth reason why it's time to bring the discussion of emotional safety into the mainstream is because emotional safety, while not talked about much, can be either a significant contributor to or detractor from a healthy culture in your organization. When employees feel emotionally safe, they are more likely to take risks, share innovative ideas, and collaborate effectively with their colleagues (Leader Factor). Emotional safety leads to better communication, innovation, and employee retention. It also helps reduce workplace stress and burnout, leading to better overall well-being for employees (Leader Factor).

For starters, let's define things for clarity.

Physical safety revolves around what people do. Most of our workplace health and safety policies are concerned with physical safety, as we put regulations, procedures, and training in place in order to enable our teams to successfully navigate work at heights, around electricity, in excavations, confined spaces, or rooftops.

Psychosocial safety revolves around what people think. We endeavor to build environments where our team members are free to express their thoughts, ideas, and opinions without the fear of being made fun of or reprised in any way. And the social element of that covers equal access to the messaging and understanding of safety by both genders and every race represented within your workplace.

This brings us to the emotional aspect. This is the aspect that revolves around how people feel. It includes a whole range of elements, not all of which can be covered in a chapter like this. But at the very least, these elements need to be called out, identified, and listed in order to enable today's safety professionals to know how to navigate this space.

Let's be honest. Diego Bellini is spot on when he observes that every organization is an emotional place because its functioning depends on human beings and the relationships between human beings (Bellini, 2022). Organizations are places of relationships and therefore necessarily elicit complex systems studded with emotions (Bellini, 2022). Every reader who has ever had a job can attest to the truthfulness of the above. To push this point further, in an effort to improve organizational performance, companies often force their employees to display emotions in conformance with a default standard which may differ from how they really feel (Bellini, 2022). This is dangerous because the expectation of this nature can create an emotional dissonance, as one can only be coerced to pretend for so long. The exact length of time that individuals can play along and what happens after the pretense is not feasible of course differs from person to person, but what is the same across the board is the fact that this strategy is not sustainable over a long haul. The state of emotional dissonance in itself can further contribute to negative thoughts, nervousness, anxiety, emotional exhaustion, and overall reduced well-being.

It is much better for individual workers and the organizations that employ them to gain a deeper "self" and "others" understanding of the emotional state, the implications of various core emotions, and their impact on the organizational culture, climate, and ultimately productivity. The aim of this chapter is to shed more light on the emotional side of safety in order to equip frontline supervisors, midlevel management, executive leadership, as well as other partners, like people or human resources, finance, reliability, risk management, etc., in understanding and operationalizing this essential element of who we are.

Among core emotions are anger, sadness, fear, and joy. Lack of knowledge or understanding in how to control or reign in one's emotions and/or when emotional health is not sound will often have a detrimental impact on one's well-being and further result in anxiety, depression, and, in worst case, suicide. We usually lump the latter under the subject of mental health but, in my opinion, a more holistic consideration allows us to consider them as part of a bigger picture of this chapter.

ANGER

"Ira brevis furor," said the Romans – anger is a temporary bout of madness. Were they right? Read on and decide for yourself.

Here is what we observe when we look around us today. We live in an angry world. And let's face it, there is plenty to be angry about. If you don't come across something on any given day that makes your blood boil, an argument could be made that you are either not paying attention to the world around you or are not in possession of strong convictions in life.

Yet, if anger is allowed to take control of your life, it will quickly eat your proverbial lunch.

Before we really dive into the sea of anger and especially its effect on workplace safety, perhaps a simple consideration is in place. Is there such a thing as righteous anger? And is there any place for it in our workplaces?

We think the simple answer to this question is "yes." Yes, there is such a thing as righteous anger. When one experiences injustice, when one witnesses unethical behavior, or when one comes across capricious and inconsistent application of rules, policies, and procedures, the resulting emotion is what Hoffman coined "empathetic anger," which is another way of saying righteous anger (Hoffman, 2008). And to be angry in situations like this is better than to turn the blind eye to them, ignore them, or pretend that nothing is wrong. Because to be angry is to realize that there is a sense of violation against what ought to be. To be angry is to do something about it that restores the right balance. Anger in this context means action. And passivity means inaction and status quo, or allowing negative actions, behaviors, and interactions to keep happening. So, as long as this empathetic or righteous anger is controlled, it is arguably a better response than one of passivity.

Now, we should move on to considering how all the other types of anger can impact safety. Anger can rob your ability to make a sound decision. It often blinds you to all the options you may have, especially if some of them are provided by the unfortunate object of your anger. Anger can lead to depression, anxiety, and self-harm.

Anger is often accompanied by the history of substance abuse. The combination of alcohol and drugs with anger often leads to rapid escalations in the workplace, at home, and in between. These rapid escalations, if not checked, often cause violent outcomes.

Anger can cause headaches and lead to problems with digestion. What's worse, anger can result in insomnia, which only exacerbates everything manifold. When left unmanaged and uncontrolled over an extended time, anger can lead to Type 2 Diabetes, which is tragic.

Anger also stresses the heart and can result in greater risk of coronary heart disease. According to research, there is a higher risk of cardiovascular events shortly after an outburst of anger (Mostofsky et al., 2014). In fact, those already suffering from arrhythmias, either in the lower or upper chambers of the heart, can see those conditions exacerbated by anger. This is because adrenaline, which increases when you are angry, can cause electrical changes in the heart (Cornwell, 2024).

Additionally, anger can disrupt digestion, and multiple studies have linked stress resulting from this negative emotion to a number of gastrointestinal complications, such as inflammatory bowel disease, irritable bowel syndrome, and gastroesophageal reflux disease. That anger contributes to poorer mental health comes as no surprise. Individuals who have difficulty controlling their anger often suffer from anxiety and depression. Also, these same people often have a hard time concentrating and their prolonged state of anger has a negative outlook on life and a harmful effect on relationships, both at home and at work.

In the complex and interconnected work environments, where my safety often depends on not only my actions and choices but also the decisions of those working next to me, it is easy to see why understanding the danger of uncontrolled anger and knowing how to reign in this emotion is so critical to workplace safety. We have all experienced or seen situations when anger caused riskier actions, responses, or reactions. It is as if all good judgment exits the scene when we allow ourselves to be overcome by anger. We literally lose control, and it is our strong negative emotion that enters the driver's seat.

But it goes beyond that. It is critical for public safety too. Think about all the road rage accidents that you have seen, heard about, or maybe even experienced yourself. It is not a stretch to say that anger kills in this context.

One thing that is different about anger is that it affects not only those who fail to control it but also those who happen to live, work, and interact with them. Anger creates a stressful environment, which impacts the health and well-being of those around. There are plentiful statistics for the negative effects of anger both on the perpetrator or the angry person and the unwilling participants who have to be exposed to the outbursts of anger simply because they work with, report to, or have to interact with the anger donor. What all the reader has to do is simply Google this information and results will pop up rapidly.

But statistics aside, we personally experienced the horrible effects of anger in various jobs that we held in our lives. Some time back, I, Simon, worked for a small outfit in Dallas, Texas, where the main office assistant was a really nice and soft-spoken middle-aged lady. She confided that this job had given her a stomach ulcer, for which she had to see her personal physician. The doctor confirmed to her after extensive testing that her condition was in fact caused by the chronic stress, which of course in this case was brought on by the owner's anger outbursts.

I, C. Thomas, can also think of many examples from my work experience where being a recipient of someone's uncontrolled anger gave me and everyone around a sour stomach, discouraged the newer and younger personnel from asking too many questions, and created an unhealthy and toxic work environment. No creativity was seen in these places. No one dared to speak up. No one dared to challenge anything. Everyone just wanted to get away from the anger donors and organizations that employed them.

At another time, I, Simon, worked in a mid-size energy service company with operations throughout the United States, whose owner would explode with anger, which resulted in a lot of yelling at anyone that happened to be around at the moment. What was even more unique and strange about this situation was that this owner's wife worked in the office with him and a lot of times the two would get into a full-blown domestic dispute with yelling and cursing and everything else – for all to hear and enjoy. This type of environment, believe it or not, is more common than we think and many if not most of us have either perpetrated or "enjoyed" the benefits of being around anger donors. And we know what this environment does to our concentration, our ability to think and make good decisions, our psychological comfort and safety in the workplace.

Workplace violence is another subfield of environmental health and safety that has seen greater attention as of late, as a result of the higher frequency of our society's inability to deal appropriately with anger. Overcome by this powerful negative emotion instead of controlling it, we are controlled by it, with the unfortunate outcome being not only broken relationships but sometimes violent outbursts against other employees, contractors, and others. This is yet another very clear way that anger can take lives.

SADNESS

Sadness is a sometimes negative and sometimes more neutral core emotion. Sadness arises from a sense that someone or something valuable has been lost (Frijda, 2005). It is a fairly common negative emotion and one of the most accurately recognized emotions by observers.

Sadness increases impatience, thus creating a potential hazard of rushed or ill-considered decision-making. When combined with the fact that sad people have been shown to be risk seeking, the recipe may produce undesirable outcomes (Spassova, 2023). Being aware of this potential trap when you are overcome by sadness or find yourself working alongside individuals overwhelmed by this emotion may help save lives, beginning with yours. It simply means that you need to be extra situationally aware, not only with regard to your actions but theirs also.

The cognizance of this risk-seeking tendency within sad individuals should inform our approach not only to the oversight of the frontline personnel particularly in high hazard situations, but also to those team members in upper management who find themselves amidst high-stake negotiations and other interactions with high impact potential. With regard to the latter point, there are research findings that suggest that sadness expressions can increase the expressers' ability to claim value in negotiations because they make recipients experience greater others-concern for the expresser (Sinaceur, 2015). I realize that this last point may fall somewhat outside the perimeters of this chapter, but it is a valid point that still needs to be made.

Samuel Taylor Coleridge (1772–1834), the English poet and philosopher, experienced profound bouts of anxiety and depression throughout his life. This life experience may have given rise to his famous phrase "a sadder and a wiser man," appearing as the ending of his 19th century *The Rime of the Ancient Mariner* poem:

> He went like one that hath been stunned,
> And is of sense forlorn:
> A sadder and a wiser man,
> He rose the morrow morn.

Interestingly, hundreds of studies within the field of psychology, performed in the 20th century, have found support for the sadder-but-wiser hypothesis: Sadness and depression make individuals wiser than non-depressed or happy people (Lerner, 2012). For example, sadness tends to be associated with careful, deliberative, *System 2* thought, as opposed to heuristic, impulsive *System 1* thought (Kahneman, 2011). So, this means that on the positive side, sadness can be characterized by the front-brain thinking, which is slower, more methodical, and thoughtful, as opposed to the more rapid and intuitive back-brain reaction.

Additionally, as related to this research, sadness has been shown to reduce a range of cognitive biases including having an overly optimistic view of one's abilities and overattributing causality to individuals. In the safety context, this translates to a more realistic perception of one's limitations and perhaps lesser proclivity to take risks that result from overconfidence. In interpersonal or work-related conflicts, individuals who are experiencing sadness are less likely to assume negative intentions and motivations by others.

While discussing the more constructive byproducts of sadness, it bears to mention that expressions of this distressing emotion make people more cooperative and willing to help. In the context of the organizational culture, sadness, while not anything that anyone would ever choose, may sometimes be a blessing in disguise by bringing about a certain level of healing to your relationships, calmness to your conflicts, and more willingness to extend the benefit of the doubt in your interactions. This observation is highly pertinent to the overall well-being of your team.

By the same token, sadness interferes with people's ability to process information. "As a result, ... sad individuals are posited to process information less systematically in judgment and decision making" (Raghunathan and Pham, 1999). So, if you know someone on your team or within your organization who lost a loved one and has gone through a relationship breakup or experienced another life-changing event that gave rise to profound sadness, it may be wise to not assign that individual to a particularly complex task or involved responsibility. Unable to process information as comprehensively or quickly as those individuals that feel good or feel neutral, your sad individuals may be more prone to errors that lead to incidents.

Observing someone in a state of sadness makes others around them share in that emotion, thus impacting the overall morale of a greater team. So, what do you do? How do you ensure that your sad teammates are safe?

To summarize, first be aware of what they are going through and of how they are perceived. Sad individuals are often deemed as weaker and could be taken advantage of. Be willing to listen, to be there, to be quiet in their presence, and to stand up for them, when the situation calls for it. More than anything sad team members may just need to be noticed and to be by their coworkers that care. Sadness, unlike anger, will usually pass if given enough time, support, and care.

Understand that their sad state affords them the ability to evaluate life with a different set of priorities and, as a result, some of their perspectives may be wiser than those of their coworkers who feel good or neutral. If not wiser, then at the very least, some of your personnel that feel sad will be more deliberate and purposeful in their decision-making. They will be utilizing the slower front brain, as opposed to the faster and more instinctual back brain mode, which is actually a good thing for your organizational safety.

Because of the reduction in biases, individuals experiencing sadness within your organization may be less likely to be engaged in conflict or escalate an ongoing argument, which is also a positive outcome. Finally, because in others sadness impedes their ability to process information and sometimes results in rushed or ill-considered decision-making, you should carefully evaluate what tasks and jobs to assign to those experiencing this distressing emotion, so as not to expose them inadvertently to more than they can handle. This could increase the risk of human error, leading to incidents and injuries.

If anger is an overwhelmingly a negative emotion that is extremely hazardous to safety, well-being, and morale of your personnel, with an adverse aftermath on your organizational culture, sadness, as shown above, has more of a mixed bag of effects and necessitates a gentler and more nuanced approach.

FEAR

Fear is a negative core emotion that is characterized by different reactions, such as anxiety, apprehension, fright, panic, and terror (Bellini, 2022). In some ways, fear is inevitable today. Why would I say that? Think about it – nearly every day, we are bombarded by news media, television shows, movies, commercials, billboards, books, magazines, social media posts, and new reports of government overreach and corruption elucidating new reasons to fear for our physical and psychological well-being. Some have even argued that we exist in a "culture of fear" (Kish-Gephart, 2009). And fear in uncertain circumstances may result in fight, flight, or freeze behaviors. This emotion and the behavior responses it activates can have a significant effect on one's psychological and physical health.

Fear can incapacitate those individuals overwhelmed by it and render them unable to act, thus leaving you, your co-workers, and your organization stranded and maybe even unable to respond to emergencies and other critical situations. To protect yourself from this type of risk exposure, it is important to foster a culture where your teams are free to express themselves on one hand and where paying attention to the expressions of coworkers is encouraged on the other hand. This act of paying attention is just plain empathy and it can play a significant role in preventing incidents, injuries, and fatalities.

By the same token, however, fear may prevent some employees from taking on greater risks and being overconfident, which may protect you from the dangers resulting from such biases. History is full of examples of significant incidents caused by overconfidence, with the sinking of Titanic, the nuclear meltdown in Chernobyl, and the more recent OceanGate's Titan implosion being just a few of them. Yet, before you begin celebrating fear, it is still important to remember that, in spite of a few seemingly positive byproducts of fear, the preponderance of research indicates that fear is generally associated with a negative effect on well-being, safety at work in terms of the number of accidents, near-misses, and behaviors related to safety at work (Bellini, 2022).

Another harmful effect of fear in organizational contexts may be that it keeps your team members from speaking up where it matters. You may have a real hazard here or a seasoned employee who takes unnecessary risks there. You may be relying on a process that is deeply flawed. But if your people are too afraid to speak up due to personal and professional considerations, to provide feedback, or to forewarn you, negative outcomes are inevitable in the well-being of the collective enterprise. And lest you think that your organization is impervious to this phenomenon, approximately 50% of over 40,000 survey respondents in a high-tech company studied by Detert (2003) did not agree that it was "safe to speak up" at work. So, if this happens in the types of companies that pride themselves in having the most open and safe cultures out there, it could definitely be happening in your organization.

Interestingly, and the limited scope of this chapter will not allow us to pursue this tangential thought as far as it needs to be pursued, the fear to speak up can sometimes be tempered or trumped by another emotion, described above, namely anger. You see, the angry person makes more optimistic risk assessments about the future and the likely success at changing the situation (Lerner and Keltner, 2000). Anger, therefore, sometimes acts as a counterweight to fear's inhibitory tendencies (Kish-Gephart et al., 2009). The interplay of these emotions and its combined effect on organizational interactions is still not clear to the scientists and would be another worthy research, perhaps reserved for a future day.

Additionally, fear encourages avoidance behavior, a narrowed perceptual and cognitive focus on perceived threats, and pessimistic judgments about risks and future outcomes (Kish-Gephart et al., 2009). This behavior can lead to missed opportunities, ill-advised expenditures, and wasted resources. In the realm of safety, avoidance behavior can be tied to taking shortcuts and making perilous judgment calls, due to inability to consider all the variables in complex decision-making scenarios.

Some fears can be within an organizational control, such as the ones originating from poor managerial styles, negative culture, unreasonable expectations, and disregard for individual personalities of team members. Other fears may stem from larger societal effects, like the state of the economy, political unrest, or global pandemics. The recent COVID-19 scare provided a perfect illustration of this point. Responding to the first type of fears may be quite straightforward, as improvements in the areas that cause them will greatly decrease or even eliminate the occurrence of this negative emotion. Attempting to normalize things within organizations when the fears are not as easy to control, however, may be a lot more challenging. The good news is that it is not impossible. The negative effects of fear on safety can still be reduced within organizations even when the causes of fear stem from outside of the organization – with the help of clear and consistent communication and social support.

To recapitulate, today's environment is fraught with fear mongering and every person around us has something to be fearful about, whether based in reality or distant potentiality. Fear can result in fight, flight, or freeze behaviors with far-reaching consequences. Fear can make your team members afraid to stand up and speak up when to do so would have been in the best interest of your organization. Fear can trigger avoidance behavior and bring about inability to act and make quick life-saving decisions. Fear can be a silent killer of your culture, productivity, innovation, and a drain on the psychological, emotional, and physical health, safety, and well-being of your team.

If you become aware of the presence of fear within your walls, it is important that you act to do everything in your power to eliminate it. If it cannot be eliminated, it must be minimized. If it is outside of your control, seek to counterbalance it with communication and support.

JOY

Joy is a positive core emotion. It is not as common as anger or fear, which is a commentary on the society in which we live, but it can be found, and it can and should be fostered. Those individuals and organizations that are fortunate enough to choose it and pursue it reap substantial and overwhelmingly positive benefits.

Perhaps it is appropriate to start with a little clarification. Joy and happiness are often used interchangeably but the two are not the same. Happiness technically refers to the pleasurable feelings that result from a situation, experience, or objects, whereas joy is a state of mind that comes from our connection with others, our family, or God and can be found even in times of grief or uncertainty (Collier, 2022). The reason this distinction needs to be made here is because we would like to establish that it is possible to cultivate joy even when the circumstances around us are far from perfect or just plain awful.

Although the neurobiology of joy is complex, there are a few neurotransmitters that stand out in promoting positive feelings: dopamine, serotonin, oxytocin, and endorphins (Collier, 2022). Laughter produced by joy, for example, releases endorphins, which is why after laughing, you feel an overall boost to your well-being and may even have a decrease in pain. Likewise, spending time with a baby releases oxytocin, the "cuddle hormone" that makes you feel connected.

Joy can boost immune system, as having a good laugh helps to lower cortisol and increase your immune cells, providing you with a higher resistance to sickness (*Laughter Is the Best*). Engaging in a good laugh helps your body relax by releasing tension and stress, which will leave your muscles relaxed for 30 minutes after you finish laughing. Go on and smile right now even if you don't have anything to smile about. Just do it, please. There, you have just tricked your brain into thinking that you are joyful, so in return you can enjoy the increased focus necessary to get through the rest of this chapter.

Studies have even shown that laughing for 15 minutes a day can burn up to 40 calories and embracing joy is a critical leg of a three-legged stool of holistic weight loss. According to Dr. Melanie Rotenberg, the author of *Laugh Yourself Thin*, one of the main reasons why most people who go on diets or pursue Ozempic-like weight loss solutions end up gaining back all the weight they lost and then some at the conclusion of their efforts is because their overall approach is more mechanical than holistic. For the diet or the pursuit of weight loss to produce permanent effects, all three legs of the stool have to be included: (1) thinking and behavior (joy); (2) input, as in food and drink calories; and (3) output, as in metabolism and activity calories (Rotenberg and Rotenberg, 2010).

Joy-induced laughter increases blood flow and blood vessels' function, which can help safeguard you against a heart attack and other cardiovascular problems that tend to show up in our senior years. Thus, joy can prolong our lives.

Joy helps build friendships and strengthen your existing relationships. So, there is a social element to it and joy is just contagious enough to where more people will want to be around you. There is a domino effect to this, and joy has a tremendously positive impact on our mental health. Laughing and engaging in things that bring you joy can help you deal with harder emotions, addressed above, like anger, sadness, and anxiety.

As is clear above, joy is more than just a fleeting emotion; it is a vital indicator of our well-being and a powerful catalyst for navigating challenges in our lives (Oaten,

2024). If the core emotion of joy is crucial to our physical, emotional, mental, psychological, and spiritual well-being, how do we pursue it? There are many different activities and ideas that could be attempted in the daily pursuit of joy. Each individual reader will have to find what works within their particular context. The list below was collected with the help of Dr. Stephanie Collier's and Jennifer Oaten's articles, entitled "How Can You Find Joy (or at Least Peace) during Difficult Times?" and "The Science of Joy," respectively.

- Connect with others – according to Harvard University's extensive research, close relationships are the cornerstone of our joy, outshining material goods, money, or fame.
- Perform regular aerobic physical activity – this can be thought of as a bubble bath of neurotransmitters, the effects of which linger long after the exercise is over.
- Put others first – volunteering and other similar selfless activities produce greater joy than focusing on self.
- Pray – we were designed to be more than just physical beings and connecting with the spiritual side through prayer fosters feelings of gratitude, compassion, and peace, all of which help build a joyful disposition.
- Find a new pursuit – we are hardwired to experience joy when experiencing novelty, so learning something new can help foster that.
- Allow yourself to rest – this will re-energize and reinvigorate your life.
- Pay attention to the good – start documenting at least three good things you notice each day and you will find yourself developing a joyous mindset.
- Limit negativity – whether it's gossipy coworker, a toxic relationship with a family member, or a complaining friend, spending time with negative people taints our perspective and robs our joy.
- Focus your efforts on what brings meaning in your life – pursue something bigger than yourself, as opposed to a bigger bank account (for me this pursuit is my Christian faith and nurturing my walk with God).
- Ask your doctor whether your medications can affect your ability to experience joy, especially if you are taking antidepressants.

Emotional culture of joy significantly and positively influences both psychological safety and relational attachment (Aboramadan and Kundi, 2023). This, in turn, improves morale, increases employee engagement, and results in safer personal choices, as well as motivation to be the brother's keeper or watching out for one another. Happier employees make safer employees. This is a perfect illustration of why human resources teams should work in a much closer fashion with their safety counterparts, as their goals are interrelated and mutually supportive.

Having touched on the types of activities that can help on to pursue joy on a personal level, the rest of this section will be dedicated to providing some tips and suggestions on how the atmosphere of joy can be erected, improved, and sustained in the workplace. Some of the suggestions and strategies provided below are sourced from the 2021 article, appropriately entitled "Creating Joy in the Workplace," authored by

Rozita Jalilianhasanpour, Shadi Asadollahi, and David M. Yousem in the *European Journal of Radiology*.

Philosophically speaking, the pursuit for a joyful workplace should begin with the right mission, vision, and values statements. This may sound overly simplistic but if the journey does not start here, then your workers will not have something bigger than themselves that they can embrace and get behind.

The second element of a multiprong pursuit of a joyful workplace could be a re-evaluation of the state of your facilities or physical plant. The physical environment of your workplace should make your employees smile by promoting cheerfulness, increasing employee engagement, and offering appealing indoor and outdoor spaces for people to gather and interact. Even something so simple as appropriate lighting, which is discussed in the next chapter, could make a real difference in fostering positive interactions. These interactions bolster employee relationships, create joy, and spark creative juices (this is especially significant, since the current post-pandemic state of employee connections at work is at all times low and was the focus of a special 2024 advisory from the U.S. Surgeon General Dr. Vivek Murthy).

Joyful interactions are a true win-win situation for the employees and the employer. The employees win in the areas of emotional, psychological, relational, and physical health and well-being. The employers win because the creative juices and ideas that such joyful interactions produce translate into decreased absenteeism, presenteeism, incidents, injuries, fatalities, while at the same time bringing about an increase in productivity, innovation, and ultimately profitability.

The third element that could contribute to a joyful workplace is encouragement of positive attitudes and discouragement of negative ones. What are some of the positive attitudes? Gratitude comes to mind first.

The attitude of gratitude can be infused into the work culture by the leadership via genuine expression of thankfulness, sincere acknowledgment of effort and achievement, and appreciation for a job well done. Interestingly, the presence of gratitude in the workplace eliminates such negative attitudes as jealousy, perception of injustice, and rancor. Gratitude not only makes for a joyful workplace but is one of the few immaterial ways that the employer can make a lasting impact in the lives of his employees because it is contagious and makes one feel better about personal life too.

Gratitude doesn't cost anything, doesn't reflect on your bottom line, and doesn't increase your payroll, so there is no excuse for not practicing it. It can be expressed in a number of creative ways: from setting aside time in group meetings to acknowledge others to recognizing employee contributions in newsletters, to rewarding an exceptional performer with a front parking spot for the month, to any other imaginative recognition. The little things you do here are guaranteed to pay tremendous dividends in the organizational culture by increasing reciprocity, cooperation, and altruism.

Another positive attitude that is sure to foster a joyful workplace is servant leadership. Since Chapter 3 already covered this area in some detail, all I will add here is to say that servant leadership can brighten your place of work, empower and lift others, and make yours the place of joy that everyone wants to be a part of.

The fourth element of a joyful workplace is the presence of humor and openness to lighten the mood when the right moment presents itself. As a non-native speaker of

English, I am well aware that fun is not a word that is present or can be translated into every language. There is no direct equivalent for it in my first language, for example. So, I will not go as far as to say: "Make sure you have fun at work." But I will say this: "Don't take yourself too seriously even at work." Do the best work you can. Be responsible and take care of yourself and others. But also, be open to laughing, even at yourself if it lightens the mood. Humor can increase cohesion in the workplace. According to Cindy Lamothe,

> Laughter releases dopamine, enhances immunity, lowers stress hormones, increases blood flow, and strengthens the heart, but beyond its many health perks, a good sense of humor leads to increased optimism, which in turn, boosts our resiliency and enables us to thrive when we're faced with adversity.

(Lamothe, 2018)

Given all the positive outcomes of joy, employees and employers alike should engage in a continual pursuit of this core emotion. Individuals need to discover what brings them joy and incorporate these activities in their daily lives. Employers and managers should be focused on creating an environment that is a pleasure to work in, attracts the best employees, retains the most productive workers, and serves as a source of joy for the people in the workplace (Jalilianhasanpour et al., 2021). By actively cultivating joy-producing pursuits in our personal and organizational contexts we can unlock a happier, healthier, and safer future for ourselves and those around us.

CONCLUSION

That we are all very emotional beings is beyond contestation. That many if not most of us could do better in understanding, managing, and leveraging our emotions is also true. The good news is that the subject of emotions and emotional intelligence or EQ has been sufficiently raised in sales and leadership literature, among others. EQ was even brought up in Chapter 3 above. But what role emotions could play in building safer cultures is something that has not been sufficiently considered, that is until now. The chapter that you just completed provided some ideas, direction for further research, and some pointers for unlocking this powerful tool within your workplace and homeplace.

Start with self-awareness of how your anger, sadness, fears, and joy can help or hurt your physical, psychological and spiritual wellbeing, as well as that of others around you. Understand and embrace the emotional side of you but do not be ruled by your emotions. Even the Bible is full of instructions of how we should rule our emotions and not be ruled by them. For example, we are commanded not to "let the sun go down" on our anger (Ephesians 4:26), to not fear (Luke 12:32), to not be anxious (Matthew 6:24–34), and to not be prideful (James 4:6), rejoice always (1 Thessalonians 5:16), and count even hard times as joy with God directing our lives. The 10th Commandment forbids covetousness, which is a sinful emotion.

Once you succeed in building this self-awareness, your next step will be becoming aware of these same core emotions in the lives of your coworkers, leaders, direct reports, family members, friends, children, spouses, and really anyone whose path you cross. Unleashing the power of self- and others-awareness will enable you to take your personal and organizational life to the next level. Be careful with this information because in the words of Spiderman, "with great power comes great responsibility."

BIBLIOGRAPHY

Aboramadan, Mohammed and Kundi, Yasir Mansoor. "Emotional Culture of Joy and Happiness at Work as a Facet of Wellbeing: A Meditation of Psychological Safety and Relational Attachment." *Personnel Review* 52(9), 2023: 2133–2152.

Bellini, Diego, Cubico, Serena, Ardolino, Piermatteo, Bonaiuto, Marino, Mascia Maria, Lidia, and Barbieri, Barbara, Understanding and Exploring the Concept of Fear, in the Work Context and Its Role in Improving Safety Performance and Reducing Well-Being in a Steady Job Insecurity Period. *Sustainability* 14(21), 2022: 14146. https://doi.org/10.3390/su142114146

Carnegie, Dale. *How to Win Friends and Influence People.* Pocket Books, 1998.

Coleridge, Samuel Taylor. *The Rime of the Ancient Mariner.* 1834. www.poetryfoundation.org/poems/43997/the-rime-of-the-ancient-mariner-text-of-1834

Collier, Stephanie. "How Can you Find Joy (or at Least Peace) during Difficult Times?" *Harvard Health Publishing,* 2022.

Comfort Keepers. "Laughter Is the Best Medicine: The Benefits of Joy. www.comfortkeepers.com/offices/minnesota/twin-cities/resources/resources/laughter-is-the-best-medicine-the-benefits-of-joy/

Cornwell, Stephanie. "5 Ways Anger Affects Your Health." *Everyday Health.* Feb. 13, 2024. www.everydayhealth.com/news/ways-anger-ruining-your-health/

de Sousa, Ronald. *The Rationality of Emotions.* Cambridge, MA: MIT Press, 1997.

Detert, J. R. *To Speak or Not to Speak: The Multi-Level Influences on Voice and Silence in Organizations.* Unpublished Doctoral Dissertation, Harvard University, 2003.

Donoghue, Kate. *Emotional Decision Making: Hardwired and Helpful.* Advocacy and Evidence Resources – Temple University Beasley School of Law. https://law.temple.edu/aer/2024/09/07/emotional-decision-making-hardwired-and-helpful/#_edn3

Emotional Safety at Work. Leader Factor. Oct. 16, 2024. www.leaderfactor.com/learn/emotional-safety-at-work

Frijda, Nico H. "Emotion Experience." *Cognition and Emotion* 19(4), 2005:473–497.

Hoffman, M. L. "Empathy and Prosocial Behavior," in M. Lewis, J. M. Haviland-Jones, and L. F. Barrett (Eds.), *Handbook of Emotions* (3rd ed.), New York: Guilford Press, 2008.

Jalilianhasanpour, Rozita, Asadollahi, Shadi, and Yousem, David M. "Creating Joy in the Workplace." *European Journal of Radiology,* 145, 2021.

Kahneman, Daniel. *Thinking Fast and Slow.* Farrar, Straus, and Giroux, NY, 2011

Kish-Gephart, Jennifer J., Detert, James R., Treviño, Linda Klebe, and Edmondson, Amy C. "Silenced by Fear: The Nature, Sources, and Consequences of Fear at Work." *Research in Organizational Behavior,* 29, 2009:163–193.

Lamothe, Cindy. The Benefits of Laughing at Yourself, According to Science. www.shondaland.com/live/a21755063/benefits-laughing-at-yourself-self-deprecation-science-psychology/, 2018 (accessed 30 November 2024).

Lerner, Jennifer S., and Keltner, Dacher. "Beyond Valence: Toward a Model of Emotion-Specific Influences on Judgment and Choice." *Cognition and Emotion,* 14(4), 2000:473–493.

Lerner, Jennifer S., Li, Ye and Webber, Elke U. "The Financial Cost of Sadness." *Psychological Science,* 24(1), 2012: 72–79. https://doi.org/10.1177/0956797612450302

Mostofsky, Elizabeth, Anne Penner, Elizabeth and Mittleman, Murray A. "Outbursts of Anger as a Trigger of Acute Cardiovascular Events: A Systematic Review and Meta-Analysis." *European Heart Journal* 35(21):1404–10. June 1, 2014.

Oaten, Jennifer. The Science of Joy. Santa Maria College. March 15, 2024. https://santamaria.wa.edu.au/science-of-joy/

Raghunathan, Rajagopal and Phamm Michael Tuan. "All Negative Moods Are Not Equal: Motivational Influences of Anxiety and Sadness on Decision Making." *Organizational Behavior and Human Decision Processes* 79(1), July 1999: 56–77.

Raghunathan, Rajagopal, Phamm Michel Tuan, and Corfmanm Kim P. "Informational Properties of Anxiety and Sadness, and Displaced Coping." *Journal of Consumer Research* 32(4), 2006: 596–601. https://academic.oup.com/jcr/article/32/4/596/1787470?login=true

Rotenberg, Melanie W. and Rotenberg, Mitch. *Laugh Yourself Thin*. Greenwood Publishing Group: Santa Barbara, CA, 2010.

Sinaceur, Marwan, Kopelman, Shirli, Vasiljevic, Dimitri, and Haag, Christophe. "Weep and Get More: When and Why Sadness Expression Is Effective in Negotiations." *Journal of Applied Psychology*, 100(6), 2015:1847–1871.

Spassova, Gerri and Palmeira, Mauricio. "Thou Shalt Be Safe: Risk Preferences in Choice for Sad Others." *Journal of Behavioral Decision Making*, 36(5). https://onlinelibrary.wiley.com/doi/full/10.1002/bdm.2350

Tyson, Lance. *The Human Sales Factor: The Human-to-Human Equation for Connecting, Persuading and Closing the Deal*. New York, NY: Morgan James Publishing, 2022.

8 The Impact of Human-Centric Lighting on Health and Productivity

Marosh Kulhavy

INTRODUCTION

How we light up our homes and workplaces has changed a lot over the course of history. Long ago, people only used sunlight, candles, or fires to see at night or inside. Later, people started using lamps that burned oil or other fuels to give light. After electric lights were invented, they became widely used to light up different spaces. These changes greatly affected how we see and feel about our homes and workplaces. They have also changed our moods, health, and how well we can do different tasks every day.

In this chapter, we will explore how human-centric lighting (HCL) can help improve our health and make us more efficient at our tasks. It helps our bodies and minds feel better. We will explore how it works, see real examples, and consider what might come next. HCL is designed to help us by copying how natural light changes throughout the day, so it meets what our bodies need. This lighting can make it easier for us to get things done at work and home. Understanding how lighting has changed over time and how HCL works can make our spaces better for our health and well-being.

EVOLUTION OF LIGHTING

The journey of lights started with natural sunlight hours, which have usually been vital in keeping our body clocks and joint health in check.

In Ancient Times, people's lives aligned with the natural light–dark cycle (light and darkness fashioned how humans slept and labored). People planned daily activities around daylight hours and worked when it was light, but they rested when it became dark.

When electric lighting was introduced, people could do tasks after darkness, which became convenient but also messed with our bodies' natural rhythms. This caused issues, like problems with sleeping and feeling down. As synthetic lighting

DOI: 10.1201/9781003583103-9

TABLE 8.1
Historical Perspectives on Lighting

Era/Times	Lighting Source	Societal Impact
Ancient Times	Natural daylight	Shaped biological rhythms and behavior of early humans.
Medieval Times	Natural daylight, firelight	Activities mirrored natural light-dark cycles; architectural construction was as per maximization of daylight.
Electric Lighting Times	Artificial lighting	Allowed expansion of active hours and more productivity but resulted in diminished natural light exposure and distorted circadian rhythms.

improved, we saw how much it influenced our fitness. This led lighting designers to create fixtures and structures that reproduce natural light to make artificial light less harmful to our bodies. Table 8.1 shows that using artificial lighting lets us work for longer and do more, but it also cuts down our time in natural light and disrupts our body clocks. HCL has been developed to bridge the gap between synthetic and natural light.

HUMAN-CENTRIC LIGHTING

HCL is designed to meet what we can and can't see by copying how daylight changes during the day. The lights we use help us see and do tasks, but they also affect things inside our bodies, like how we feel, our mood, and how awake we are. HCL differs from regular lights because it does more than help us see. It also looks at how light affects our bodies, like our internal clocks, our feelings, and how awake we are. The right kind of light can help us think better, sleep better, and feel happier by focusing on how light impacts us in ways we can't see.

Therefore, in order to have an HCL approach, we should start from the analysis of human needs (Figure 8.1).

Also, HCL can save energy by ingeniously using light during the day. By making artificial light work more like natural light, HCL can improve our health and help us get more done at home and work. This type of lighting makes our living and working spaces better and helps improve our overall quality of life.

A schematic diagram in Figure 8.2 shows pathways that light takes in the brain, and corresponding responses – both the visual (solid path [B]) and nonvisual (dashed path [A]) response. The visual pathway ends in the visual center and the nonvisual path works its way into the body's functions via the suprachiasmatic nucleus (SCN) which is biological or master clock and other areas involved in the regulation of circadian rhythms. Table 8.2 points out these pathways, showing that the way we see things lets us feel stuff, while another way helps keep our body clocks and hormones in check.

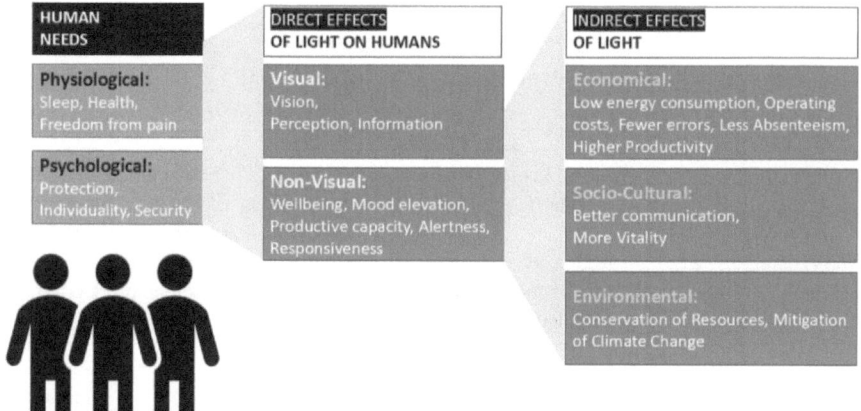

FIGURE 8.1 Human needs and effects of light.

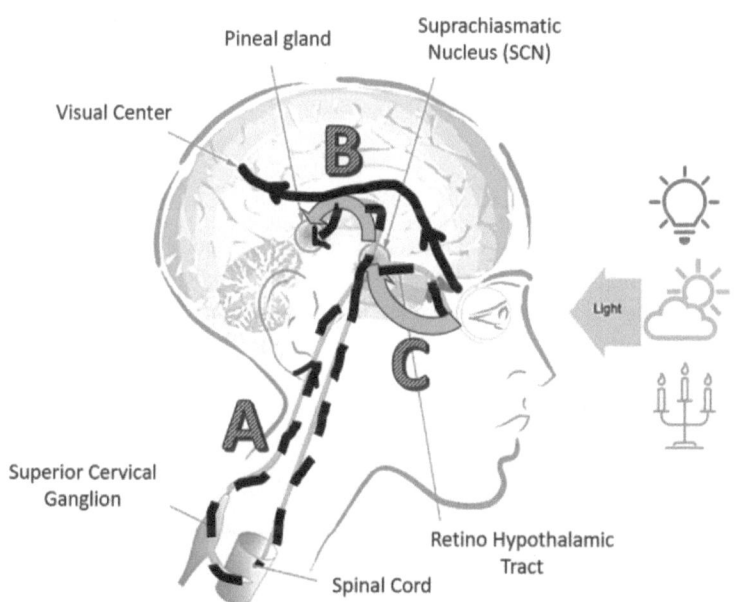

FIGURE 8.2 Effects of light – visual and nonvisual pathways.

VISUAL EFFECTS OF LIGHT

Our everyday lives depend on light because it allows us to see and complete our tasks. The retina contains photoreceptors. When light reaches our eyes, these photoreceptors capture it. They then send signals to our brains, enabling us to see clearly. Good light helps us see better, makes our eyes feel less tired, and allows us to finish our tasks

TABLE 8.2
Visual and Non-Visual Control Mechanisms of Light

Mechanism	Pathway	Function
Visual Pathways (solid path [B])	Eyes to Primary Visual Center/Cortex	The ability to perceive images, colors, and motion, enabling interactions with the environment.
NonVisual Pathways (dashed path [A])	ipRGCs to Suprachiasmatic Nucleus	Regulates circadian rhythms via sleep–wake cycles, hormone production, and health in general.

well. Good lighting is also essential because it keeps us safe, lets us work better, and makes us feel more comfortable, especially in places like schools and offices.

- **How We See:** The main job of our visual system is to help us see clearly and move safely through our surroundings.
- **Doing Our Tasks:** When the lighting is good, we can do our tasks better, with fewer mistakes and less strain on our eyes.
- **Eye Care:** Too dark or bright light can tire our eyes, which can affect their health.

NONVISUAL EFFECTS OF LIGHT

Light helps us see and affects other essential parts of our bodies. These effects happen in ways that do not help with vision but instead control how our bodies work internally. Some special cells in our eyes called intrinsically photosensitive retinal ganglion cells (ipRGCs) send signals to a part of the brain known as the SCN. The SCN acts as the body's master clock and controls our circadian rhythms. Circadian rhythms are the body's internal 24-hour cycles influenced by light and darkness. These cycles control things like when we feel sleepy and when our bodies make hormones and other essential functions. Getting the right light at the correct time is crucial to stay healthy.

- **Circadian Rhythms:** Light plays a significant role in controlling our body's circadian rhythms. These rhythms affect how well we sleep, hormone production within our bodies , and our body temperature.
- **ipRGCs and Biological Clocks:** ipRGCs play a crucial role in sending signals to the SCN, which facilitates the synchronization of the brain's internal clock with its surroundings.
- **Health Implications:** Disruption of circadian rhythms, frequently due to inappropriate lighting, can lead to health troubles such as sleep problems, mood modifications, and reduced cognitive performance.

Light is the main Zeitgeber (time cue, from the German for "time givers") of human life because it establishes our circadian rhythms: when light is off and when it is

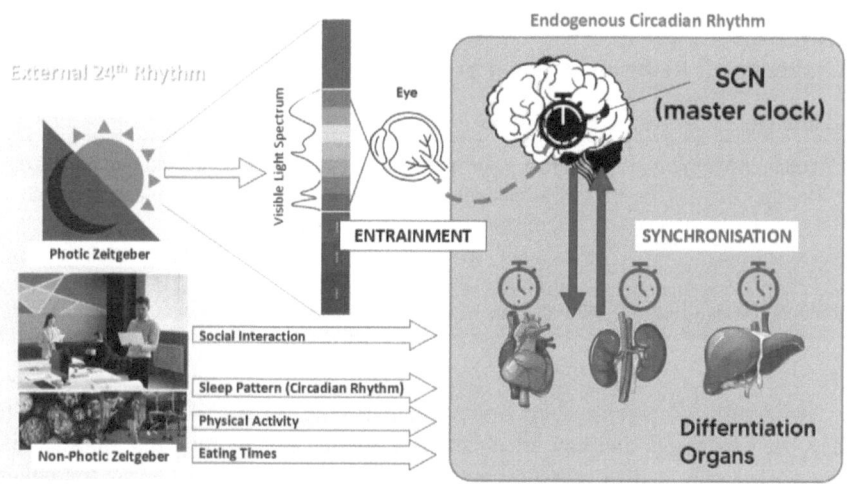

FIGURE 8.3 External 24-hour rhythm and circadian synchronization.

on. Light is the major Zeitgeber that synchronizes the biological/master clock in the SCN. Other Zeitgebers, such as eating, social interactions, and physical activity, are also important for good entrainment (Figure 8.3).

When entrainment is "good," this means the internal clock is well aligned with the 24-hour day–night cycle. Hence, physiological and mental health is at its best, sleep quality improves, and overall well-being is achieved. Proper entrainment means physiological processes follow predictable patterns aligned with time of day, which is critical for sustaining alertness, mood, and general health.

The disorders of circadian misalignment are due to disruptions in entrainment related to artificial sources of light, which generates irregularity and therefore predisposes to pathologies like sleep disorders, depression, and chronic diseases.

HCL is meant to help our body's natural systems by providing the right kind of light in right places. This allows people to maintain their best health and well-being. By thinking about what we see and how light affects us in other ways, HCL can improve our happiness with life and make us healthier overall.

HEALTH IMPACTS OF HUMAN-CENTRIC LIGHTING

Good natural or artificial light is essential for maintaining the body's health. Light helps us control our body's internal clock, which lets us know when to sleep and when to awake. If this clock gets out of order, it can cause trouble with sleeping, make us feel grumpy, create health problems, or make us feel sick quickly. Lights designed to match our biological body rhythms can help prevent these problems. For instance, bright, blue-toned light in the morning can help us experience greater alertness and think better. At night, warm light helps our bodies loosen up and prepares us for sleep. Blue light in the daytime helps us stay awake, but too much blue light at night can make it hard to sleep. This is why having the right kind of light at the right time is very important.

TABLE 8.3
Challenges of Reduced Daylight Exposure

Challenge	Description
Circadian Rhythm Disruption	Inability to be exposed to natural light treads internal clocks out of alignment, thus negatively impacting sleep and hormone regulation.
Mood Disorders	Not enough daylight is associated with increased risks of depression and anxiety.
Reduced Productivity	Not enough daylight may have adverse impacts on productivity.

A healthy and regular sleep is vital for our mind because it affects how we feel, how well we sleep, and how much stress we have. Bad lighting that doesn't match our wishes can lead to pressure, tension, and tiredness, affecting our fundamental well-being. As indicated in Table 8.3, less time in daylight can mess up our body clocks, make us feel down, and drop our work output. Special lighting that changes throughout the day can assist us in feeling much less stressed, sleeping better, and being in a better mood. This can also make us more productive and happier. Schools that use this lighting have observed that kids focus better and behave better. Nursing homes have additionally seen that people sleep better and feel calmer.

CIRCADIAN RHYTHMS AND HEALTH

As stated before, circadian rhythms depend on the 24-hour light and dark cycle and are essential for the health of the body and mind (Figure 8.4). When these rhythms are out of sync, it can cause sleep problems, sadness, problems with thinking, and even heart issues. Shown in Table 8.4, when our body clocks are off, it can lead to big health problems, like sleep issues, feeling sad, body problems, heart issues, and even a higher chance of some cancers. Human-focused lighting works by giving the right light at the right time. It helps people sleep better, keeps their minds sharp, and makes them feel good. Bright light during the day, especially in the morning, helps the body's natural clock stay in line with day and night. This allows people feel more awake and alert and keeps their natural rhythms from getting off track. Keeping this balance is vital for doing well during the day, having steady hormones, and staying in a good mood.

It is also essential to stay away from blue light at night because doing so allows the body to make melatonin, which is the hormone that helps people sleep better. Blue light late at night can stop melatonin from working, which makes it harder to fall asleep, lowers the amount of sleep, and causes tiredness the next day. Warm light in the evening, like the colors of the setting sun, tells the body it's time to relax. This helps people move into sleep more smoothly and makes their sleep better. Human-centric lighting is helpful to health by ensuring lighting is set up to keep the body's timing right. This helps people stay healthy both right away and over time.

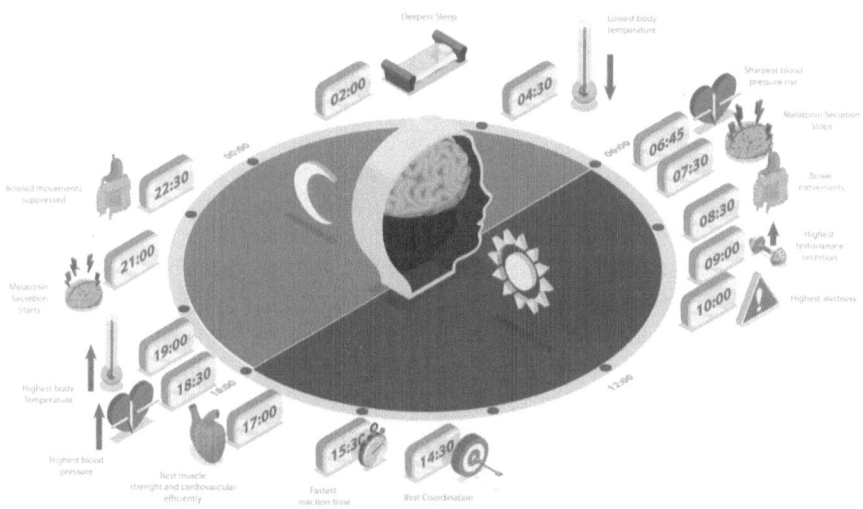

FIGURE 8.4 Twenty-four-hour biological clock – circadian rhythm, natural cycle for healthy sleep and routine. (Licensed Image: AdobeStock_705337173.png.)

TABLE 8.4
Potential Health Issues from Circadian Disruption

Health Issue	Description
Sleep Disorders	Insomnia, fragmented sleep, and poor sleep quality due to misaligned sleep–wake cycles
Depression	Abnormal light exposure promotes mood disorders.
Metabolic Diseases	Obesity, diabetes, and metabolic syndrome due to increased risk.
Cardiovascular Disease	The malfunction of the circadian rhythms can cause a high risk of heart conditions.
Cancer	There is a higher incidence of certain types of cancers associated with chronic circadian disruption.

Also, people working at night or having mixed-up schedules can get unique HCL setups to help them handle their misaligned rhythms. The right light at correct times can ease some of the harmful effects of night work, like not getting enough sleep or feeling fuzzy mentally, which leads to better health and more productivity.

PRODUCTIVITY BENEFITS OF HUMAN-CENTRIC LIGHTING

Statistics provided by the University of Cambridge reveal that 54% of individuals say their work environment adversely affects their well-being, and in many offices up to 60% of employees do not have enough access to daylight. These issues can greatly influence employee health and productivity. Good lighting is important for how well people can work, especially in offices. Lighting that focuses on people and is set up the right way can help workers feel good, lower tiredness, and improve how well their minds work. For example, a study by the Fraunhofer Institute found that workplaces with HCL had a 10% increase in how well people worked and a 15% drop in people missing work.

Offices with HCL also have workers who are happier with their jobs, miss less work, and do better overall. Lights that change during the day can help workers feel more awake, stay focused, and feel less stressed, which makes the workplace better in the long run. As more companies understand how important it is for workers to stay healthy and work well, HCL has become crucial to how workplaces are designed today, improving how happy and well workers perform.

Custom lights and knowing about different types of sleep can make people work better.

PERSONALIZED LIGHTING AND CHRONOTYPES

PERSONALIZED LIGHTING

Lights matching each person's sleep pattern, whether morning or evening, can help them work better and get more things done. Bright, blue-enriched light for morning types boosted early-day performance, helping them live alertly and energized all through the morning hours. Evening sorts gain from gradually increasing light intensity, which permits them to ease into a productive state as the day progresses. This approach could be more effectively aligned with their natural biological rhythms; however, it also reduces the chance of fatigue and improves overall mood. Personalized lighting can also mitigate the harmful effects of mistimed light exposure, ensuring people keep ideal energy levels and awareness. Using unique lighting solutions, businesses can create workspaces that meet different needs. This helps make each person work better and improves how groups work together. Having personal lighting also makes a workplace where everyone can work well, leading to good results for workers and their bosses.

CHRONOTYPES

Chronotypes describe a person's habitual inclination regarding the instances of the day they choose to sleep or be active. There are extraordinary chronotypes.

- **Morning Types (Larks):** Tend to awaken early and feel most effective in the morning.
- **Evening Types (Owls):** Feel more fantastic, alert, and efficient later in the day.
- **Intermediate Types:** Some people are more than just morning or night people. They are a mix and do their best work during the middle of the day.

Understanding a person's daily schedule, called their chronotype, is essential. This helps you choose the best lights for your workplace/offices. When light suits someone's natural rhythm, it can help them work better and be more productive.

There is also a difference between a person's natural body rhythm and the schedule they follow in their daily life. The *biological chronotype* is a person's natural time for sleeping and being active, while outside things like work hours or family duties influence the *lifestyle chronotype*. When these two chronotypes don't match, it is often called social jetlag, which can make people tired and lower how well they work.

Social jetlag is very common today because work and social responsibilities may not fit someone's natural sleep and wake times. Understanding how important it is to match your natural sleep schedule with your lifestyle can help improve wake-up times, how well you think, how active you are, and when you go to bed. For example, people who naturally like staying up late might try waking up earlier or using light with more blue in the morning to feel better and have less social jetlag. When we understand these differences, human-focused lights can make places where people do their best work. By giving different lights to different people, we can help them feel more in tune with their body clocks. This can lower stress and improve how well they do their tasks and how much they like their jobs.

CASE STUDIES AND REAL-WORLD APPLICATIONS

CORPORATE IMPLEMENTATIONS

Some workplaces now use HCL systems to help workers stay healthy and do their jobs better. For instance, in Amsterdam, The Edge (15-story office building, known for being smart and sustainable) has a Philips Lighting system. It changes how bright the light is and the color at different times of the day. This makes it more like the light outside. This helps workers feel more awake and keeps them in a better mood. It also helps save energy. Good lighting like this can make workers feel better and work harder, keeping them active and motivated. This copies the way light works in nature. Workers said they felt more awake and in a better mood. They also saved energy. These systems show how innovative lighting plans can make workers happier and more excited to work, helping them stay fit and motivated.

Another example is Deutsche Bank's headquarters in Frankfurt, where HCL was installed to cope with various lighting needs throughout various workplace regions. The system decreased employee fatigue, absenteeism, and power intake while increasing productivity. By customizing the lighting fixtures for exceptional styles of workspaces, including meeting rooms, open workplace areas, and personal offices, Deutsche Bank ensured that employees had been supplied with the most fulfilling lighting for their particular obligations. This level of customization helped foster a more cozy and effective work environment, ultimately benefiting each employee and the organization.

ARCHITECTURAL CONSIDERATIONS

Including HCL in building designs needs careful planning. Architects must consider where lights and windows go to ensure people get the right amount of natural and artificial light. This helps meet human needs. The design should support body rhythms and make it easy to see clearly. Windows should be placed to let in as much natural light as possible. Artificial lights should work with natural light to keep things balanced during the day. Architects and lighting designers should work together to ensure the lights work well and look good. This planning helps people feel more comfortable, healthy, and connected to their surroundings.

FUTURE DIRECTIONS AND INNOVATIONS

The future of human-centered lighting is based on using facts to make lighting personal. New devices that sense light and machine learning help change lights to fit each person's needs. These changes ensure people get the right light to feel good and stay healthy all day. This way, lighting can be made to help people do their best and feel their best. By using innovative technology, HCL will keep improving at helping people in their daily lives. The focus is on making lighting work for each person in the best way possible. Wearable gadgets can check non-public light exposure and provide real-time remarks to change light conditions mechanically. Such technology can lead to even more personalization and effectiveness of HCL, as lights may be continually optimized primarily based on each individual's necessities.

HCL fixtures are likewise covered in broader health programs that integrate workouts, proper vitamins, and stress control. By combining those unique elements of health, the excellent outcomes of HCL are accelerated, leading to higher average health and well-being. For instance, using HCL to support the proper circadian alignment can enhance sleep quality, improve energy levels, and encourage participation in physical activities. This complete method increases the sound effects of HCL, making it an essential part of plans to help people feel better and be healthier.

New ideas may bring HCL together with smart home systems and building designs. These intelligent systems will adjust lighting based on the time of day, the weather, or what people do. This will make lighting easier to control and more helpful. Experts are also looking into new lighting ideas like tunable LED lights and better materials that spread light. Natural sunlight could make lighting work better and feel more comfortable. Combining HCL with new technology has a lot of possibilities. As our understanding of HCL grows, we will see even more ways it can help with health, wellness, and saving energy.

CONCLUSION

Lighting is essential for keeping our bodies work well and helping people stay healthy and do their best. From the sun's natural light to new modern lights, HCL shows how much we know about how light helps us feel better physically and mentally. Good lighting can change how we live every day. It can help us sleep better, feel happier, alter our hormones, and stay healthy. Over time, we have moved from just using sunlight to using HCL systems. This helps us understand more about how light can make us feel better and how we can adjust to our surroundings. HCL can make places healthier and better for working, supporting how our bodies naturally work. HCL can make people perform better at work, sleep better, and feel good overall. Putting HCL in schools, offices, and homes can give people the right light at the right time. This helps them think more clearly, feel more emotionally balanced, and stay healthier. HCL does not just help people, it can also save energy and help the planet by using light more efficiently. By learning more about how light and health are connected, we can make better choices to create spaces that help people feel and work better.

BIBLIOGRAPHY

1. Kulhavy, M. (n.d.). *Beyond Just Light: The Role of Human-Centric Lighting in Enhancing Data Center Operations*. DataCenterDynamics. Retrieved from www.dat acenterdynamics.com/en/opinions/beyond-just-light-the-role-of-human-centric-light ing-in-enhancing-data-center-operations/
2. Kulhavy, M. (n.d.). *The Crucial Role of Human Factors in Data Centers*. DataCenterDynamics. Retrieved from www.datacenterdynamics.com/en/opinions/ the-crucial-role-of-human-factors-in-data-centers/
3. Boyce, P. R. (2014). *Human Factors in Lighting* (3rd ed.). CRC Press.
4. Rea, M. S. (Ed.). (2010). *The IESNA Lighting Handbook: Reference and Application* (10th ed.). Illuminating Engineering Society.
5. Veitch, J. A., & Galasiu, A. D. (2012). *The Physiological and Psychological Effects of Lighting and Daylight*. National Research Council of Canada.
6. Gottlieb, J. F., Benedetti, F., Geoffroy, P. A., Henriksen, T. E. G., Lam, R. W., Murray, G., Phelps, J., Sit, D., Swartz, H. A., Crowe, M., Etain, B., Frank, E., Goel, N., Haarman, B. C. M., Inder, M., Kallestad, H., Kim, S. J., Martiny, K., Meesters, Y., Porter, R., Riemersma-van der Lek, R. F., Ritter, P. S., Schulte, P. F. J., Scott, J., Wu, J. C., Yu, X., & Chen, S. (2019). The chronotherapeutic treatment of bipolar disorders: A systematic review and practice recommendations from the ISBD task force on chrono-therapy and chronobiology. *Bipolar Disorders, 21*(8), 741–773.
7. Signify. (2017). *Connected office lighting – The Edge*. Retrieved from www.assets. signify.com/is/content/PhilipsConsumer/PDFDownloads/United%20States/ODLI201 71026_001_UPD_en_US_PLt-1637CS_Edge_case_study_June2017.pdf
8. Mario Bellini Architects. (2010). *"Green Towers" – Headquarters Deutsche Bank*. Retrieved from www.german-architects.com/en/mario-bellini-architects-milano/proj ect/green-towers-headquarters-deutsche-bank
9. LightingEurope. (2015). *Quantified Benefits of Human Centric Lighting*. Retrieved from https://www.lightingeurope.org/news-publications/publications/153-quantified-benefits-of-human-centric-lighting-april-2015
10. van Bommel, W. (2023). Human-Centric Lighting. In *Encyclopedia of Color Science and Technology* (pp. 907–910). Springer. https://link.springer.com/referenceworken try/10.1007/978-3-030-89862-5_426
11. Czeisler, C. A., & Gooley, J. J. (2007). "Sleep and Circadian Rhythms in Humans." *Cold Spring Harbor Symposia on Quantitative Biology*, 72, 579–597. DOI: 10.1101/ sqb.2007.72.064
12. Stevens, R. G., & Rea, M. S. (2001). "Light in the Built Environment: Potential Health Effects of Chronic Circadian Disruption." *Cancer Causes & Control*, 12(3), 279–287.

9 Operationalizing Neurodiversity and Human Performance to Enhance Safety and Business Success

Randy Cadieux and Noah Cadieux

INTRODUCTION

This chapter explores an emerging area of interest among the safety community: neurodivergent (ND) workers and how they may help advance organizational success. The chapter provides background information and a framework for integrating neurodiversity into the workplace, and one point that must be emphasized is the research nature of this chapter. This work explores areas that, to date, seem to be under-researched. There is little research on neurodiverse workers and risk exposure/mitigation outside of traditional office or home office settings. The concepts in this book are exploratory, where the authors apply concepts from the fields of safety, human performance (HP), ergonomics, and design to neurodiversity and develop theoretical frameworks for risk mitigation and performance enhancement.

It is the belief of the authors that this information has the potential to accelerate individual, team, and organizational performance, but this is in no way guaranteeing a level of safety or performance. This chapter creates a foundation, and additional research-in-practice will be needed to validate the concepts described in this book. However, this is an important topic and the authors' hope is that this book will foster further conversations around neurodiversity in all industries and that industries will start incorporating modern approaches toward safety, wellness, and HP to create alignment between ND workers and organizational policy and management attitudes. Industries and organizations wishing to advance their performance may be well-suited for neurodiversity hiring programs, and the concepts in this chapter may help by providing a "blueprint" for integration and implementation strategies.

DOI: 10.1201/9781003583103-10

BACKGROUND

Modern safety science is a highly interdisciplinary field drawing from several academic disciplines, with the field of psychology being vital to understanding how people think and function within a workplace environment. Often the study of psychology focuses on the properties and functions of the "typical" or "average" person as a baseline for human behavior and capabilities. Designs of engineered systems are most often based on the characteristics of the "average person" or the 80% of the population that falls under the center of the bell curve of any given psychological trait (Fung, 2021). Significant deviations in brain structure and function or behaviors which are deemed abnormal by our culture are typically classified as disorders within documents such as the *Diagnostic and Statistical Manual of Mental Disorders* (DSM) or International Classification of Diseases (ICD).

For several decades, there has been significant criticism of this pathologization and othering of people with neurological conditions. Throughout the history of the DSM's development and revisions, many social scientists and philosophers questioned its role in pathologizing of "deviance," which they argue may be representative of human's variability rather than conditions that truly impair and reduce well-being of those who have them (Kawa & Giordano, 2012). This critique laid the groundwork for the idea that many neurological conditions could be thought of as a form of natural human variation, and that physical and social environments may cause some to experience significant disability rather than neurological differences being innately disabling. This would lead to the development of the neurodiversity paradigm, with sociologist Judy Singer (1998) being credited with introducing the concept in academia in her dissertation describing how online communities of people with autism thrived through shared online spaces in the late 1990s.

This concept spread from the field of sociology and disability studies to psychology and neurology, where some researchers have begun to consider neurodevelopmental conditions to be natural forms of human neurological diversity (Fung, 2021). One specific form of the neurodiversity paradigm is the Strength-Based Model of Neurodiversity, which considers that changes in brain structure and function seen in neurodevelopmental conditions such as autism, attention deficit hyperactivity disorder (ADHD) and dyslexia may result in domains of both cognitive strengths and challenges, with the ultimate focus being placed on enhancing strengths rather than mitigating biological deficits (Fung, 2021).

The interest in neurodiversity has been growing over the past several years, with numerous organizations recognizing not only the challenges neurodivergent workers face in the workplace, but also emphasizing the strengths these workers can bring to the organization. Many organizations are recognizing the many strengths neurodivergent workers bring to employers, depending on their condition and individual skill sets, including collaboration, problem solving, focus, and innovation. We would be remiss not to mention Elon Musk, whose self-attested neurodiversity has certainly not limited or impeded his ability to innovate, perform, and progress in his multiple fields of expertise. Moreover, Musk's status as the richest man in the world could make for an argument that his "condition" may actually be somewhat of a strength...

While there is a growing movement to create more and more organizational neurodiversity hiring programs, and Employee Resource Groups (ERG), many of these programs appear to be primarily focused on work performed in a traditional office or work-from-home office environments. This is a great step forward, and its importance should not be understated, but it is likely that there are numerous neurodivergent workers (who either have a diagnosis or who possess a neurodiverse condition but have yet to be diagnosed) who would prefer non-office jobs that typically involve exposure to engineered systems, such as heavy machinery and equipment where physical hazards must be mitigated, or who may already be working in these high-risk environments.

There seems to be a lack of emphasis on placing neurodivergent workers in positions in industries where they may be exposed to physical hazards, and there is little research into how certain symptoms of neurodiverse conditions may combine with other hazards in the operational environment to create unique risks that may not exist to the same extent for neurotypical workers. To a large extent, this is understandable because for many neurodivergent workers, depending on their condition(s), these work environments may not be suitable and may present challenges that exceed their ability to cope in these environments. On the other hand, there may be numerous neurodivergent people already in the workforce or who are seeking employment who may be highly interested and capable of performing work in positions outside of a traditional office setting. Additionally, organizations that primarily conduct their operations in environments other than traditional office buildings and home offices may be missing tremendous opportunities for accelerating their performance by not creating neurodiversity programs that allow for placement of neurodivergent workers in these roles.

This chapter will discuss the implications of the neurodiversity paradigm on the fields of safety and human and organizational performance (HOP). It will lay the groundwork for explaining how organizations can align opportunities for current or potential neurodivergent workers to advance organizational performance through a step-wise process of operationalizing neurodiversity that includes going beyond basic safety measures and moves toward advancing worker wellness that may also positively impact morale, retention, and team performance. This process includes the use of HP principles, health and wellness focused programs, and several international and national standards around safety, health and wellness to unlock new areas of individual, team and organizational performance. While not providing an exhaustive list of neurodivergent conditions, this chapter will introduce conditions which have been well researched from a neurodiversity perspective and are often focused within organizational neurodiversity initiatives.

Autism Spectrum Conditions

Within the DSM-5, Autism Spectrum Disorder (herein referred to as ASD) is defined as a neurodevelopmental condition resulting in deficits in social communication as well as restricted and repetitive patterns of behavior and interests (DSM-5 TR, 2022). Alterations in sensory processing, including both hypersensitivity and hyposensitivity, to stimuli also frequently occur within ASDs, and fine motor skill deficits may also occur within ASDs (DSM-5 TR, 2022). ASDs may also co-occur with intellectual disability, although the majority of those diagnosed with ASDs have average or above average IQ scores (DSM-5 TR, 2022). These combinations of difficulties are a particular challenge for adults with ASDs seeking employment, with unemployment of adults with ASDs being estimated to be as high as 85% (Krzeminska et al., 2019). ASDs are also associated with significant strengths, particularly in domains of non-verbal intelligence, long-term memory and attention to detail (Fung, 2021). ASDs are also associated with participation in science, technology, engineering, and mathematics occupations (Ruzich et al., 2015). Some scholars such as Judy Singer (1998) and Simon-Baron Cohen (2022) have highlighted the role of ASD traits in the development of science and technology throughout history.

ATTENTION DEFICIT HYPERACTIVITY DISORDER

As the name suggests, ADHD is defined by a combination of inattention and hyperactivity symptoms. Notably, one may be diagnosed with ADHD with only inattention symptoms or only hyperactivity symptoms, although most have a combination of both symptom sets (DSM-5 TR, 2022). Some symptoms of ADHD include, but are not limited to, difficulty remaining still, difficulty sustaining attention on a task over long periods of time, tendency to break or violate established rules and frequently changing topics during conversation or difficulty remaining on a single train of thought (DSM-5 TR, 2022). These symptoms are related to low levels of dopamine within the brains of individuals with ADHD, leading to behavioral compensation that increases this low dopamine level (Hutson & Hutson, 2022). These thought and behavior patterns may result in increased impulsivity and risk-taking behavior. On the positive side, research has also shown that these same processes are related to enhanced creativity, divergent thinking and enhanced leadership skills particularly in crisis situations where acting quickly is vital to avoid negative outcomes (Fung, 2021).

DYSLEXIA

Dyslexia falls under the DSM-5 category of specified learning disorders. Dyslexia is defined by difficulty with reading and writing when other forms of learning and intelligence are not impaired. While dyslexia is clinically defined by reading and writing ability, at the neurological level greater changes in brain regions not associated with language are also affected in those with dyslexia. The sensory cortexes in the brains of those with dyslexia have been found to be more interconnected than those without, causing the brain to process more peripheral information which may cause confusion when reading written words that require greater focus on the center of visual attention (Fung, 2021). This more "gestalt" or holistic form of sensory processing appears to play a role in enhancing creativity, divergent thinking and storytelling abilities in those with dyslexia (Fung, 2021).

Developmental Coordination Disorder (DCD)

DCD, also known as dyspraxia, is diagnosed as a lifelong difficulty with coordination and gross motor movements. Symptoms can include abnormal gait when walking, difficulty with balance, muscle weakness and fine motor skill deficits such as when handwriting or using tools. DCD also frequently co-occurs with conditions such as ADHD and learning disabilities such as dyslexia. Therefore, considering ergonomics and accessibility is important across a variety of neurotypes. DCD has not received as much attention in research from the neurodiversity paradigm as the previously mentioned neurotypes, but some research has shown that those with DCD often have strengths in social skills, verbal communication and empathy (Doyle, 2020). Currently this is assumed to be due to differences in lived experiences and problem-solving strategies developed by those with DCD in navigating life with a significant disability, but further research may potentially point to a brain-based explanation of these strengths as well.

Strength-Based Model of Neurodiversity

Moving forward with the biodiversity theory of neurodiversity and a litany of evidence supporting strengths across a wide variety of neurodevelopmental conditions, Stanford professor Lawrence Fung founded the Stanford Neurodiversity Project and the Strength-Based Model of Neurodiversity (SBMN), which "is a formulation of approaches to maximize the potential of neurodiverse people based on their strengths and interests" (Fung, 2021, p. 7). Under the SBMN, neurodiversity is defined as "the diversity that views differences in brain function and behavior as normal variations in the human population" (Fung, 2021, p. 7). Importantly, the SBMN acknowledges that neurodivergent conditions can produce disabilities and challenges, and whether they produce challenges or strengths is dependent upon contextual factors (Fung, 2021, p. 7). In particular, this model highlights the role of brain structure and function in producing strengths.

Reframe Model

Another model for neurodiversity with a focus on workplace and coaching application is the reframe model coined by industrial/organizational psychologists Nancy Doyle and McAlmuth (2023). The authors of this model argue that what is considered to be a symptom of disability in one context may be a strength in another context, and highlight the role of the workplace environment in producing potential disability of workers. As noted earlier, those with ASD tend to have difficulties breaking from routine and may feel distress when routines are disrupted (DSM-5 TR, 2022), but this may cause those with ASD to be particularly attuned to drift from procedures or other small changes within the environment that increase the risk of injury. Accidents frequently occur due to operators not noticing a crucial change in environmental conditions and not changing their procedures or behavior accordingly (Pupulidy &

Vessel, 2023). Therefore, having the capability to detect these changes is vital for ensuring safety.

BIOPSYCHOSOCIAL MODEL OF NEURODIVERSITY

To understand why a particular condition can have strengths, challenges or potentially a combination depending on the context, it is important to consider how disability is framed. A traditionally medical model of disability assumes that the condition itself is what impairs and disables the individual, whereas a more contemporary social model of disability would highlight how a lack of consideration and accommodation of varying levels of ability by society is what produces disability in individuals with medical conditions (Whelpley et al., 2023). Individual psychology can also play a role in disability, where two individuals with similar symptoms and social context may experience different levels of impairment depending on factors such as resilience, motivation and pride (Whelpley et al., 2023). Individually, each of these models may not address all factors that contribute to disability, which is why many experts in the fields of neurodiversity advocate for a biopsychosocial model of disability and neurodiversity. Under the biopsychosocial model, there are clear biological indicators of "spiky profiles" where neurodivergent individuals excel in some areas and struggle with others, and these neurobiological spiky profiles can be accommodated by adjusting the fit between the individual's abilities and the working environment (Doyle, 2020, pp. 112–114).

NEURODIVERSITY AND SAFETY

SAFETY DIFFERENTLY AND HUMAN PERFORMANCE PRINCIPLES

The biopsychosocial model of neurodiversity and SBMN demonstrate how neurodiversity can yield both strengths and disabilities depending on environmental context rather than ND individuals simply being "disordered." This older view of the individual being inherently wrong or broken was also present for much of the safety industry's lifespan. Traditional views of safety posit that many accidents are caused by human error, where a worker violated a rule, performed a task incorrectly, or omitted steps in a task leading to an incident. According to Conklin, the traditional safety philosophy argues:

> Workers are the problem to be fixed. We fix safety by making workers better.
> We must tell workers what to do and, perhaps more importantly, what not to do.
> Safety is the absence of accidents.
>
> **(2019, p. 20)**

Akin to how the neurodiversity paradigm seeks to enhance agency of ND individuals through focusing on strengths and interests rather than deficits, prominent safety professionals began to argue against "fixing" the worker and the idea that safety is something to be "done to the worker" (Conklin, 2019, p. 17). Sidney Decker was the first to coin the term "Safety Differently" (Conklin, 2019, p. 19), from which Conklin adapts the four principles of Safety Differently:

1. Safety is not defined by the absence of accidents, but by the presence of capacity.
2. Workers aren't the problem, workers are the problem solvers.
3. We don't constrain workers in order to create safety, we ask workers what they need to do work safely, reliably, and productively.
4. Safety doesn't prevent bad things from happening, safety ensures good things happen while workers do work in complex and adaptive environments (Conklin, 2019, p. 24).

Aligning with these principles of Safety Differently, Conklin devised his five principles of HP:

1. Error is normal, even the best people make mistakes.
2. Blame fixes nothing.
3. Learning and improving are vital, learning is deliberate.
4. Context influences behavior, systems drive outcomes.
5. How you respond to failure matters, how leaders act and respond counts (Conklin, 2019, p. 24).

Of note, throughout this chapter the reader will find descriptions of HP and HOP. HOP is an extension of HP that includes benefits to the organization as well. In some cases, the terms will be used interchangeably, as HP principles not only benefit individual workers, but also benefit the entire organization. We believe that there is a great potential for alignment between the principles of Safety Differently and HP principles and the neurodiversity paradigm. Dekker and Conklin's work highlights the importance of worker's input and perspectives, and that they do not simply execute job tasks and introduce error into a system, but actively complete systems and create improvements through their capacity for innovation and adapting to system failures (Conklin, 2019). Regardless of how much effort is put into designing a "perfect" or optimal system, due to various resource constraints and external factors, trade-offs are required during system design and development.

Ultimately, when a system is commissioned, there will generally be suboptimal parts of the system, rendering it an "incomplete" system from the mindset of the workers. In the course of daily work, Safety Differently would not view workers who violate procedures as errant components of a perfect system, but instead view workers as those who must make adjustment and local adaptations in order to "complete" the design by filling the gaps that were not addressed during the design and development of the systems they are using. Similarly, ND workers may use their strengths to complete the design in unique ways. Not only do ND individuals possess individual strengths they can add to a system through neurological differences compared to their neurotypical (NT) peers, but they also provide different perspectives from both their unique neurological processing of their physical environment and lived experiences.

By taking in a broader set of inputs from a neurodivergent workforce, the entire organization's systems can be made more robust against error and leverage the broader cognitive diversity of the workforce for enhanced innovation and performance.

HEALTH AND WELLNESS-FOCUSED PROGRAMS

In the early stages of traditional safety, the primary focus has understandably been to reduce fatalities and serious injuries. However, hazards that cause direct physical harm are not the only potential risks that can exist within an organization. Workplace stress and fatigue, bullying and harassment are just few of the many "psychosocial hazards" that can exist within any organization (ANSI & ASSP, 2019, p. 3). As the safety progression matured, there became an increasing recognition that safety professionals must address more than just the physical risks that exist in organization. In 2003, the National Institute of Occupational Safety and Health started the Steps to a Healthier US Workforce Initiative which sought to enhance overall worker well-being through health enhancing programs and health supporting work environments which would be formalized into the *Total Worker Health®* (TWH)[1] model in 2011. TWH encompasses many areas of safety, health and well-being, including mitigating physical hazards workers may be exposed to as a foundational and critical component, but also moves beyond basic incident prevention strategies, and views workers from a holistic viewpoint, aiming to reduce risk in areas that may impact mental and emotional well-being. TWH has the potential to not only reduce negative events (like accidents and incidents) but also enhance positive factors (employee well-being, job satisfaction and worker performance).

Integrating health and wellness-focused programs will be vital in enhancing the well-being of a neurodivergent workforce. Some conditions may directly affect risk of physical injury in high-risk environments, such as motor coordination deficits seen in DCD and autism, or motor tics seen in Tourette's syndrome. Other conditions may yield increased risk of physical injury if the environment does not accommodate cognitive differences. For example, an organization relying solely on warning signage as a hazard control on hazardous equipment may increase the risk for workers with dyslexia, as the signage may be difficult for them to read. Without a secondary layer of control this signage becomes a weak single-point failure as a risk control. Across multiple neurotypes, ND workers may be more vulnerable to overstimulation and workplace bullying (Lazerwits et al., 2022: Mellifont, 2020) and be more prone to engaging in self-medication with hazardous substances that further reduces well-being (Posada & Herbert, 2022). By utilizing holistic approaches to enhancing worker well-being through a combination of risk mitigation and health and wellness enhancing strategies, organizations may be better equipped to address the particular needs of ND workers while advancing overall safety practices for the entire workforce.

Up to this point, this chapter has explained the fundamental concepts around neurodiversity, how ND workers face challenges in accommodating environments and possess strengths which have the potential to improve organizational performance, and how Safety Differently, HP principles, and health and wellness-focused programs may help create the framework for empowering ND workers for success, enabling organizations with a new way of improving operational performance. The next section will describe implementation strategies which will be required to change the ND paradigm, and fully integrate ND workers into the organization in a seamless fashion, rather than as a "bolt-on" strategy to improve performance.

IMPLEMENTATION STRATEGIES

When leaders attempt to introduce new initiatives in their organizations, without deliberate and thoughtful planning, those initiatives may be more likely to fail. As John Kotter states,

> Major change efforts have helped some organizations adapt significantly to shifting conditions, have improved the competitive standing of others, and have positioned a few for a far better future. But in too many situations the improvements have been disappointing, and the carnage has been appalling, with wasted resources and burned-out, scared, or frustrated employees. To some degree, the downside of change is inevitable. Whenever human communities are forced to adjust to shifting conditions, pain is ever present. But a significant amount of the waste and anguish we've witnessed in the past decade is avoidable.

(Kotter, 2012, pp. 3–4)

Some of the reasons for failure during transformational initiatives include too much complacency, failing to create a strong leadership team to oversee the transformation, underestimating the importance of a strong vision and under communicating that vision, allowing the transformation to be blocked by obstacles, failing to create short-term wins, declaring victory too early during the transformation and failing to anchor the changes in the organizational culture to achieve lasting change (Kotter, 2012, pp. 4–15). Fortunately, Kotter has proposed a transformational roadmap for change, which, when bolstered through the implementation of Safety Differently concepts and HP principles, may provide organizations with an advantage when trying to implement neurodiversity initiatives and programs.

In order to successfully implement organizational neurodiversity initiatives and programs it is important to consider the need for a strong foundation. Safety Differently and HP principles provide several foundational elements upon which organizations may build ND programs and initiatives. By laying the foundation of Safety Differently and HP principles organizations may already be primed to focus on creating error-tolerant systems, avoiding blaming individuals for errors and failures, examining context around work, creating learning and improvement strategies and tactics, and educating leaders on appropriate responses to success and failure. These concepts may help smooth the transition into enabling and empowering ND workers to do their best work and propel the organization into new areas of creativity and innovation.

SAFETY DIFFERENTLY AND HUMAN PERFORMANCE PRINCIPLES AS A FOUNDATION AND ND-ENABLER

As previously described, Safety Differently views workers as the source of positive capacity, able to further organizational performance by enhancing what goes right, rather than focusing on the prevention of errors and mistakes. Instead of humans being seen as problems to be solved, Safety Differently views humans as problem solvers and innovators. HP principles serve like a compass with a True North, where the principles serve as a guiding pointer for making decisions and taking action in the absence of prescriptive guidance.

The HP principles also provide a solid foundation upon which neurodiversity initiatives may be built because these principles provide the baseline guidance for supporting people in the organization, creating the conditions for psychological safety, and emphasizing learning and improving and the leadership soft skills that are required to help all workers (ND and NT) thrive. Safety Differently and HP principles may be viewed as a framework for priming the organizational culture so that it is ready to accept the uniqueness that ND workers bring to the organization.

Many organizational approaches to safety seek to protect workers from harm by, at a minimum, complying with government regulations, and, for companies with advanced safety performance goals, seeking conformance with various consensus standards or achieving certification for certain ISO standards, such as ISO 45001 Occupational Health and Safety Management System. Approaches for advancing beyond basic protection of workers and creating a safe work environment toward a holistic approach that not only protects employees, but also seeks to improve their overall wellness, which in turn enables improved performance, morale, job satisfaction and retention, may be a way to provide organizations with a competitive advantage.

SUPPORTING PILLARS FOR NEURODIVERSITY

After integrating Safety Differently mindsets and HP principles, organizations may build upon those foundational elements by creating pillars to support neurodiversity initiatives and programs. Similar to a building, HP principles serve as a foundation and the pillars serve as supports to enable neurodiverse workers to do their best work. By adopting these pillars, organizations may be able to design their work environments to be more neuro-inclusive and accommodating while not excluding neurotypical workers. Concepts such as Universal Design and Prevention through Design can guide management so that the workplace may be designed for safety from the beginning, during the design or redesign process, rather than as a "bolt-on" strategy after equipment is commissioned. Using these approaches, good design principles will benefit all workers (both ND and NT). Additional support pillars include the use of workplace self-profile tools which workers may use to determine if they possess any neurodiverse traits, and to identify their strengths and challenges which, in turn, may help them and their managers develop accommodations and risk controls to enhance their ability to perform their work while mitigating risk. For ND workers there may be cases where implementing the identified accommodations may be viewed as risk controls because these accommodations may actually work to reduce risk for ND workers, depending on their conditions.

The following section will provide a stepwise approach as a model for implementing this type of change, using HP principles as a foundation and other areas, such as Universal Design, profile software tools, training neurodiversity champions, as pillars for enabling performance. Implementation strategies following John Kotter's transformational change model will be described, including an explanation of the necessary strategic and tactical leadership involvement to integrate neurodiversity into the fabric of the organization. This transformational change model will be used to describe the stepwise approach, from creating a strong belief in the need for change and moving through strategy development, setting and achieving short-term goals, and finally working to make neurodiversity part of the organizational cultural norms, to the point where all worker types deeply believe that neurodiversity is a required part of their organization and necessary for success.

IMPLEMENTATION STRATEGIES

John Kotter proposes an eight-stage transformational model designed to take an organization from one state to a completely different state where cultural norms have shifted and been engrained in the organizational leadership and workforce. The eight stages include:

1. Establishing a Sense of Urgency
2. Creating a Guiding Coalition
3. Developing a Vision and Strategy
4. Communicating the Change Vision
5. Empowering Employees for Broad-Based Action
6. Generating Short-Term Wins
7. Consolidating Gains and Producing More Change
8. Anchoring New Approaches in the Culture (Kotter, 2012, p. 21).

Figure 9.1 shows a visual representation of the eight stages of change using HOP principles as a foundation and pillars to support the change.

ESTABLISHING A SENSE OF URGENCY

Establishing urgency is the first step in the transformational change process. The necessity and urgency for change may be driven from the bottom-up or top-down but must have leadership support at the highest levels of the organization. Senior leadership must understand the critical need for change and be able to thoughtfully and deliberately articulate the reason for change and the cost of staying the same. It must be clear throughout the organization that the time to change is now and stagnation will lead to some type of detrimental situation, such as loss of business, damage to reputation, or being displaced by a competitor. A sense of urgency may not simply be related to avoidance of a negative situation but could also be related to seeking out a positive situation, such as capitalizing on changing market conditions or customer demands, and the opportunity to create innovative products to gain a competitive advantage.

From the standpoint of neurodiversity, the urgency spans multiple areas, including the ability to innovate and solve problems to gain a competitive advantage, the ability to excel in key areas based on the unique skillsets possessed by various neurodiverse

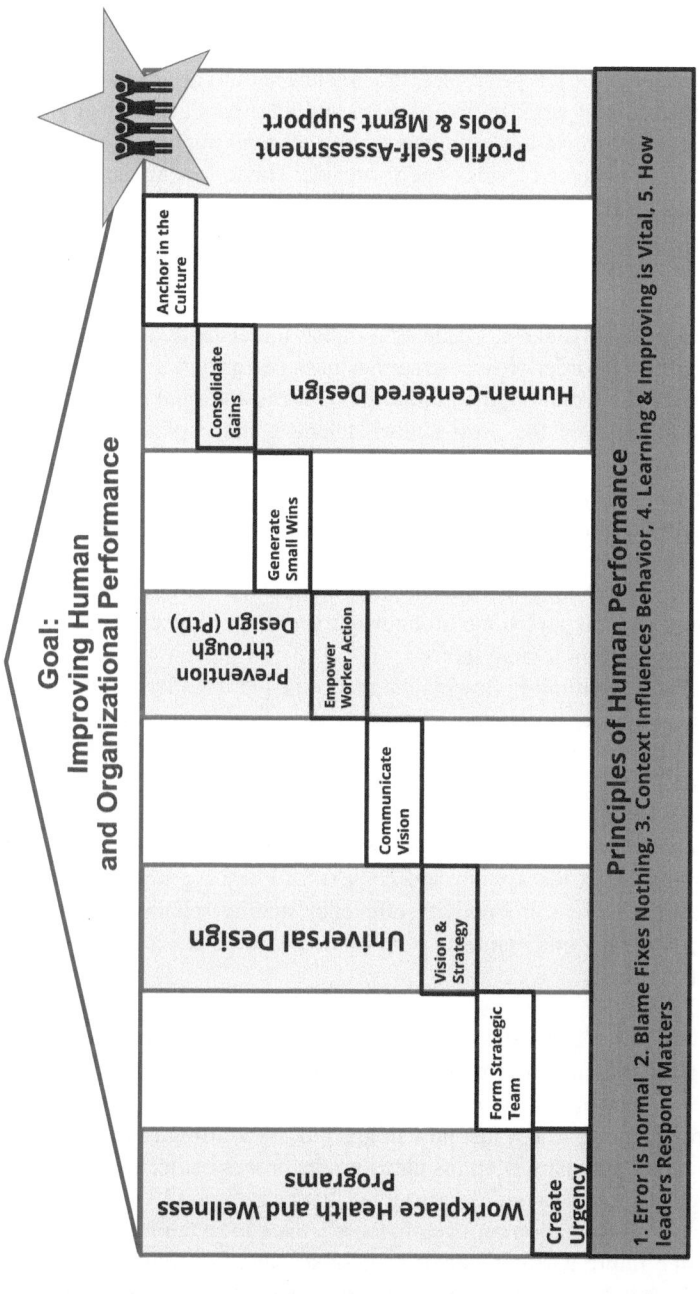

FIGURE 9.1   Transformational journey framework using HOP and neurodiversity (Cadieux & Cadieux, 2024).

workers. In order to successfully move to this desired future state, it will be necessary to create a foundation of psychological safety so that everyone in the organization could feel safe to contribute to the conversations, including sharing good news and bad news, without fear of reprisal.

From the standpoint of HP principles, the urgency should be based on improving the changes of successful work by focusing on building work capacity and adaptive capacity so that the organization is able to rapidly respond to change and taking the organization toward advanced levels of performance. The following list explains how the five principles of HP may be used to assist in generating urgency for change and how leadership can explain their importance and relationship to the desired future end-state.

- People Make Mistakes: Leadership must understand and articulate the message that in order to accelerate business performance there needs to be a mindset shift around human fallibility. Error is a normal part of the human condition and even the most skilled, talented and experienced workers at all levels of the organization will make mistakes. By adopting this mindset shift the organization can be more primed to accept that ND workers will make mistakes in unique ways, but those mistakes need not be detrimental to the business nor are they much different than neurotypical workers making mistakes. By adopting the mindset that mistakes should be seen as learning opportunities, this lays some groundwork for all employees to feel a level of support and psychological safety.

- Blame Fixes Nothing: Following the first principle if leadership is driving the urgency around accepting that error is a normal part of work, the logical argument for disrupting the paradigm of blaming workers for mistakes naturally follows. If we agree that error is normal, we can agree that blaming workers for errors and mistakes will not help us to learn and improve as an organization. Leadership should make the case that shifting this mindset from blaming to learning about the system and context will pay big dividends down the road in areas such as productivity, efficiency, quality, reliability, safety, profitability, and many other areas that are often measured using Key Performance Indicators (KPIs).

- Learning and Improving Are Vital, Learning Is Deliberate: Leadership must make the case that the risk of stagnating is greater than the effort and future reward for moving toward becoming a learning organization. Learning is more than identifying failure. It includes curiosity about normal work and how things go right, not just how things fail. By shifting to this mindset, the organization may gain deep insights into the processes, techniques and tribal knowledge from experienced resources that help to make work successful in most cases. Information captured in these areas can be fed forward to improve and refine future performance. On the other hand, learning from failure can be a very powerful tool for understanding gaps in systems, equipment, work instructions, training and other areas. But discussing these items is not sufficient for learning. To truly learn something must be done with the information, so leadership needs to drive the point home that learning needs to be a deliberate activity that is built into routine and non-routine work, such as a post-job debrief, capturing and sharing lessons learned and action items, and tracking those action items through to closure. This should include follow-up reviews to determine if those closed out action items have improved performance and not made it harder for resources to do their best work.

• Context Influences Behavior, Systems Drive Outcomes: The urgency should be around moving from human error as a cause of failure toward examining systems, processes, work methods, rules and policies to determine the influencing factors that went into decision-making processes and their potential contribution to errors and failures, as well as successes. By staying focused on human error and blaming workers for failure, learning about system gaps will be hindered and it will be more likely that these systemic influencing factors will remain as error precursors.

• How You Respond to Failure Matters, How Leaders Act and Respond Counts: Leaders' and managers' response techniques must align with the other principles, and knee-jerk reactions where employees are blamed for errors will no longer be tolerated by the organization. Staying stuck in past behavior, where leaders and managers may overreact and hasten to pass judgment and blame, will lead to organizational stagnation around safety and other key performance areas. Leaders and managers must reframe their behavior and thoughtfully respond through inquiry, curiosity and support when an error or incident occurs. They must be interested in understanding the context and system factors that influenced workers' decisions rather than being interested in who they can blame. Leaders must also emphasize positive responses to successful work accomplishments where teams are recognized and rewarded and leaders seek to understand the factors that led to success, so they may be repeated in the future.

CREATING A GUIDING COALITION

A strong steering team must be established to lead and guide the transformation. This guiding coalition will serve as the driving team for implementing the change initiatives. While they may not perform the full implementation duties, they will be required to help create and align the vision and connect the vision across the organization. The Guiding Coalition should be staffed with a good mix of those with leadership skills and management skills, and should include representatives from various stakeholder organizations, and in particular groups who will be impacted by the change and who will be required to help implement the change. This could include representation from Human Resources, Safety, Operations, Maintenance, Quality, Sales, Finance, and other key departments. Members should include a mix of senior personnel and less experienced personnel, and cognitive diversity should be sought out when determining the composition of the coalition. All members must be highly interested in the transformation and be willing to contribute.

CREATING A VISION AND STRATEGY

In this stage of change, senior leadership along with the Guiding Coalition should be working on a vision for the future. A vision may be thought of as an expression of the ultimate desire for the change which the organization should strive to achieve. It isn't necessarily an objective or measurable goal, but might be an emotional representation of the future. Kotter describes effective visions as being imaginable, desirable, feasible, focused, flexible and communicable (Kotter, 2012, p. 73). A strong vision statement should be easy to explain and understand and should serve to motivate others toward change. In the case of the neurodiversity and HP journey, the vision should reflect a future state where the organization empowers all workers for success regardless of neurotype. The vision should reflect the foundation of HP principles that will create a climate of psychological safety where the organization is focused on continual learning and improving and inclusive of all neurotypes.

The strategy should be a high-level description of the steps to achieve the vision. In the case of this specific transformational journey, the strategy should include laying the foundation of HP principles and pillars for supporting neurodiversity initiatives. The strategy development process should include a review of the remaining stages of change around communication, empowerment, achieving key milestones, building upon the change and putting systems in place to help ensure the cultural change endures. Elements of the strategy should include planning for:

- Systems and policy Review and Development for Ways to Incorporate Elements of Safety Differently and HP Principles. These could include revamping root cause analysis and incident investigations to require the avoidance of blaming individuals and teams while avoiding analysis of system and contextual factors, creating organizational policy statements that address human error and psychological safety, strategies for creating learning opportunities, and providing soft skills training for management around appropriate response to failure.
- Human and Organizational Performance Workshops: HOP workshops will be designed to educate all levels of the organization around the five HP principles and how they work to enhance organizational performance.
- HOP Champions Training and Development: This is an extension to the HOP workshops, but is dedicated to developing HOP champions, who will go through immersive training in HP principles and HOP and will be key leaders who will spearhead HOP implementation and provide support to various departments and business units.
- Neurodiversity Awareness Workshops: These workshops are designed to help all levels of the organization understand the strengths and challenges of ND workers and the value they bring to the organization, while also describing some of the unique needs each ND worker may have in order to be highly productive and effective in their roles.
- Neurodiversity Champions Workshops: This is an extension of the Neurodiversity Awareness workshops but is designed to specifically train and

develop key leaders and liaisons within the organization who take on a collateral duty of advocating for the needs of ND workers and become a critical link between various departments, including human resources, managers and individual ND workers. With more specific training on how to understand the unique challenges and opportunities for ND workers, ND champions can serve as an important resource to help with specific accommodations, as well as universal design ideas which could benefit all workers. ND champions can also help bridge the conversations between individual ND workers, other team members and managers. ND champions will not replace managers and will not impede good communications between workers and managers but will serve as a value-added resource. Building ND champions training into the strategy will be an important step.

- Health and Wellness-Focused Programs: Programs designed to protect the physical safety of workers, while also recognizing the importance of physical and mental health, and overall well-being are important for reducing risks across multiple fronts. These types of programs support the holistic worker, going beyond the workplace and recognizing that worker performance is impacted by factors on and off the job. These types of programs could benefit all workers while providing unique opportunities to assist ND workers, due to their emphasis on the total person, including mental wellness, not only the protection from physical hazards.

- Universal Design: This is a concept that emphasizes good design for all. Universal design may be simplified with the saying "what is good for one is good for many." Rather than designing the workplace for only most of a workplace population, universal design can help incorporate best practices and good practices that will benefit ND workers while also enhancing the workplace for NT workers. Examples could include comfortable lighting, reducing extraneous noise exposure through the selection of material and equipment, and placement of workstations, policies designed to help provide clear language that is easily interpretable by ND and NT workers, and accessibility features that make it easier for all workers to do their best work.

- Human-Centered Design: This approach to the design of the organization, equipment, tools, policies, procedures and work instructions, among other areas, places humans (i.e., users) at the center of the design process where solutions are focused on humans, so they are empowered to do their best work. It involves the use of empathy and overcoming the idea that there is only a single best solution (cognitive fixedness) so that solutions may be rapidly prototyped and tested to determine feasibility and applicability toward solving a problem. This agile approach is repeated in an iterative fashion until a final solution is implemented (Landry, 2020). By using the human-centered design approaches, it provides valuable insights into the challenges and solutions to help improve the chances of successful implementation of the solution.

- Work Profile Software and Management Support: Leadership should include a strategy that incorporates the use of tools to help ND workers identify their unique strengths and challenges, and specific accommodations that may be required to help them do their best work. This could include work profile soft-ware that provides feedback to ND workers, and which may be used by ND Champions and managers to identify needed accommodations.

COMMUNICATING THE CHANGE VISION

Communicating the vision is a critical stage and can either make or break the trans-formation. It is important to identify the various communication mediums and formats and create a consistent message that will resonate with everyone impacted by the change. When developing communications, it is critical for leadership to recog-nize the potential for resistance to change and for people to ask what is in it for them? Employees want to know why the change is important and why they should care about the change. Describing the vision and how it will benefit all workers, regard-less of job title or rank, will help to diffuse resistance to change. The importance of repetition should not be underestimated. Many change efforts have failed because the vision and strategy were under-communicated. In many cases, there may have been strong momentum at the start of the transformational journey, but the commu-nication frequency and content fizzled out over time, leaving employees wondering what happened, and possibly assuming the change was simply a "flavor of the month" which they can wait out until it is gone completely.

Planning for communication is an important part of this stage. Leadership should determine what written and verbal communications will look like. This could include virtual town hall meetings, road shows where senior leadership visits various business units and site locations, screensavers, emails, instant messaging channels, podcasts and other forms. The communication should include a description of the vision of a future where HOP is embedded into the organizational culture, creating a culture of learning and improving where employees are seen as problem-solvers rather than problems to be solved. Layering upon the HOP foundation, the message should clearly explain the benefits of improving neuroinclusivity in the organization and how neurodiversity programs and initiatives will serve to benefit the entire organization. Empathy is an important element of communication, and the message should also serve to put minds at ease for the upcoming transformation, so employees understand there will be minimal (if any) negative impacts, and the majority of the change will serve to benefit all workers across the organization.

EMPOWERING EMPLOYEES FOR BROAD-BASED ACTION

During this stage, workers at multiple levels in the organization will be engaged to help support the organization's transformation. Empowerment is more than a buzzword and must be more than lip service. Empowering workers means removing barriers and blockers to the change. If there are rules that inhibit or prevent change, they must be changed or eliminated. If there is too much bureaucratic drag in the organ-ization that inhibits and slows down the change, management must determine how the bureaucratic red tape may be reduced or removed altogether. It will be extremely frustrating and will erode trust if leadership says employees will be empowered to implement the change, but nothing is done to remove or reduce the barriers to change. An additional way through which workers may be empowered is through direct participation in the change efforts. Key areas include:

- HOP Champions: Managers should query their departments and seek volunteers for HOP champions training. While these individuals do not need to be supervisors or managers, they should demonstrate the ability and willingness to be a leader. HOP champions will receive in-depth training on HOP and should be given specific training on how to coach individuals and teams around HOP so they are equipped to help move beyond a blame culture to a learning and improving culture. HOP champions will serve as key resources who are used as matrixed resources to assist their own divisions/business units as well as others, depending on where they are needed. They could participate in learning opportunities and incident investigations to help provide coaching on how HOP principles may be applied for learning and improving, and helping teams avoid blaming workers for mistakes.
- ND Champions: Neurodiversity champions will serve in a similar capacity as HOP champions, serving as key liaisons and experts, but their focus will be more on coaching around neurodiversity and less on coaching around HOP. They may also be HOP champions, but in this capacity their focus will be on advocating for ND workers and serving as a key liaison between ND workers, management, human resources, employee resource groups, and other key departments and stakeholders in the organization. ND champions will receive specialized training, beyond basic knowledge on neurodiversity conditions, which could include neurodiverse conditions, including strengths and challenges, specific accommodations that may benefit workers based on neurotypes and specific conditions, training on work profile software so they can educate workers and managers on the software implementation and use cases, and coaching techniques which may be used in various situations. ND champions work as key liaisons between operational workers, HR and management. This isn't absolving management from a direct relationship with ND subordinates, as that relationship should always be respected, but ND champions will be trained on a unique skill set around working with and helping to communicate some of the challenges faced by ND workers to management and in turn convey feedback from management to ND workers. ND champions are facilitators, not managers, of the process. ND champions' work could be a collateral duty in addition to regular duties based on position, or if the organization is large enough, perhaps the ND champions could be dedicated in their role and not perform any other duties.
- Health and Wellness-Focused Surveys: As a way of empowering workers to contribute to the transformation around worker performance and wellness, health and wellness surveys may be used to gauge workers' perceptions of their own well-being. These surveys could focus on areas such as the design of the workplace and the physical work environment, the workplace experience,

organizational policies, and worker's perception of their health (NIOSH, 2024). The results of the surveys may be used to understand the workplace culture and areas where improvements may be made.

- Neurodiversity Culture Surveys: Capitalizing on the momentum gained through the health and wellness focused surveys, a more specific survey around neurodiversity in the workplace may be administered to those who are interested. The results of the survey will help reveal the level of maturity of neuroinclusivity in the organization.
- Involvement in Universal Design and Human-Centered Design Initiatives: Involving multiple stakeholders in the design process, focusing on making the workplace and job methods more neuro-inclusive will provide direct user feedback and help leadership to create more neuro-inclusive designs. Examples could include workspace designs around lighting and noise, amounts of natural light, wall colors, and other physical areas which could impact wellness and create opportunities for enhancing worker performance for all employees (ND and NT workers). Human-centered design methods can use empathetic approaches to help understand worker pain points, and move beyond cognitive fixedness, which can be a bias toward a single solution as opposed to the idea that there are multiple ways to solve problems. This iterative approach to design can provide rapid prototyping for testing and evaluation and should include a mix of neurodivergent and neurotypical workers.
- Work Profile Software: These applications are designed to help workers identify their strengths and challenges in the workplace and can help them gain an idea of how these lean toward one or more ND conditions. Based on the questions and worker-provided responses, the software can recommend accommodations which could help improve worker performance. The results of the analysis could be used by management to understand the unique strengths and challenges faced by each employee and to understand universal design approaches and specific accommodations needed by individuals. This implementation must be supported and backed by human resources to provide workers with the needed protections and psychological safety so they may feel comfortable sharing the results. Additionally, this software should be made available to all workers, including those without a diagnosis of an ND condition. This will help provide a fair and equitable approach and will help ensure opportunities exist for all workers for needed accommodations to help them thrive in the workplace. This is also important because there may be workers in the organization whose tendencies lean toward one or more ND conditions, although they may not possess a diagnosis and may not even realize that they lean toward these conditions.
- Lessons-Learned and Feedback System: Learning should be included as a deliberate part of work methods and processes. Learning is more than discussing failures, but includes learning from success and failures, capturing the information on what went well, what went wrong and recommendations for improvement, and tracking improvement action items through closure

and communication across the organization. Lessons learned is more than identifying information. Learning takes place when the information is shared and acted upon rather than simply being saved and stored on a network storage drive.

GENERATING SHORT-TERM WINS

In order to track progress along the transformational journey, it is important to identify goals or targets to achieve along the way. This should be done during the early planning stages and should be reviewed and updated in an iterative fashion as new information is learned and as plans are refined. Short-term wins may be both qualitative and quantitative, and they provide the entire organization with a means of showing progress. The wins will be evidence that the change is more than a flavor of the month. The short-term wins should be planned based on the strategy developed during stage 2 of the transformational journey and which is refined throughout the process.

Key areas and potential metrics could include:

- HOP Champions: Some of the short-term wins can include the number of HOP champions trained and a description of improvements that have been realized as a result of HOP champions being integrated into organizational operations and processes. A specific example could include a story of how a HOP champion facilitated a learning review of operational success (or failure), where context and system improvements were identified, and where lessons learned were captured, shared and implemented.
- ND Champions: Some of the short-term wins can include the number of ND champions trained and a description of the improvements that have been realized as a result of ND champions being integrated into organizational operations and processes. While there needs to be professional discretion around individuals and needed accommodations, it could be very beneficial to the organization to capture and share general information around the ways ND champions are liaising with individual workers and managers and how changes to workplace design have resulted in organizational improvements, such as innovation and creativity, productivity, efficiency, quality, reliability, profitability or other key performance areas. This may be challenging to identify directly and may take some effort, but part of the success should be captured in a qualitative manner using deep storytelling rather than simply relying on a specific KPI. For example, if ND champions have facilitated discussions around the need for design improvements related to physical workspaces, which have resulted in improved lighting, which has further resulted in increased productivity and worker happiness (which may be measured through feedback surveys), this demonstrates the value ND champions are bringing to the organization. This also helps to justify the investment in ND champions training programs.
- Health and Wellness-Focused Programs: The results of the health and wellness-focused surveys may be used to gauge workers' well-being. Information

gathered around the surveyed topic areas may help to understand the workplace culture and areas where improvements may be made. Improvements in these areas may be documented and shared across the organization. These improvements will likely benefit all workers (ND and NT). Additionally, examples of how initiatives were implemented to reduce risk through redesign of the workplace environment, and education around neurodiversity, can demonstrate the improvements in worker wellness and productivity. This will be helpful to show how the transformational processes are benefiting workers.

- Neurodiversity Culture Surveys: Survey results may be analyzed, and general information may be shared with the organization regarding accessibility, neuroinclusivity and how the organization is performing in these areas. Key leaders should use the feedback to make changes to move toward a more neuro-inclusive organization. These changes could include job design and work instruction design that includes clear language, inclusion of ND and NT workers in lessons-learned sessions, channels of communication that allow ND workers to request accommodations and recommend design changes, universal design approaches such as lighting and noise reduction, stress reduction initiatives, and other areas that demonstrate leadership commitment to creating a work environment where ND workers feel respected and where they can thrive, which will further enhance the mission of the organization.

- Results of Universal Design and Human-Centered Design Initiatives: When design initiatives are undertaken, the results should be shared across the organization to demonstrate how these approaches to problem solving were used to address challenges in the workplace that were creating barriers to improved performance.

- Work Profile Software: The organization will need to make investments in resources to procure and implement this software. It will be important to demonstrate the value and benefits of this software and how it has helped all workers, whether they have a diagnosed neurodiverse condition or not. The organization is not allowed to require workers to disclose any medical condition, so it is important to ensure this software is available to all workers. Some of the benefits may include recommendations for accommodation even for workers who may not have a formal diagnosis yet who may have tendencies towards one or more ND conditions. For example, based on the answers to the software questions it may be revealed to NT workers that they have tendencies like someone with ADHD. The recommended accommodation could go a long way in helping them improve their performance. Changes to the workplace because of the software recommendations may be described in aggregate so as to not disclose information about any individuals.

- Lessons-Learned and Feedback System: The results of lessons learned that were implemented should be shared across the organization to demonstrate the benefits of the process. Implementing lessons learned through the creation of action items, tracking action items through closure (such as through implementation of hazard controls or improvements to workplace design, tools, training and procedures) and explaining how the process resulted in improvements will demonstrate the value of the transformation. To truly integrate lessons learned in the organization will take an investment of time and other resources, so it will be important to show the efforts were worth it.

CONSOLIDATING GAINS AND PRODUCING MORE CHANGE

In this stage it will be important to realize that the transformation is not complete. It may be tempting to ease off the accelerator and assume the transformational journey may be placed on autopilot, but this could be a dangerous assumption. Kotter describes the dangers of declaring victory too early. During this stage the guiding coalition and key leaders who are helping to lead the transformation will identify new approaches to integrating the new methods into the organizational framework. This is a way of solidifying the gains achieved thus far and which have been communicated through short-term wins.

As an example, if pilot programs were trialed, the feedback may be incorporated into future implementations and the pilot programs may be transitioned into full-scale implementation. More short-term wins should be identified and shared across the organization. This stage should be used to plan for future long-term integration and should be a means of solidifying the efforts, so they move from a change process to a new state of the organization. At this point, any doubts about the change efforts being temporary should start to disappear as the guiding coalition prepares for complete integration of the changes into the rewards structure in the organization. Using this approach, the new methods will be tied to performance and rewards, so workers understand the new way of operating is the norm and there is no going backward because everyone senses the rewards associated with the new methods.

ANCHORING NEW APPROACHES IN THE CULTURE

This final stage will cement the change in the organization. At this point one of the goals should be a recognition by all employees that the new way of doing things is the right way of doing things. At this stage workers should recognize there is no going back and, in fact, they are so pleased with the new methods that they value the new ways over the old ways and have no desire to go backward. During this stage it will be important to create ways to onboard new employees into the culture and create plans to handle naysayers who have not gotten on board with the new vision. There may be some people in the organization who need to leave if they do not support the new culture around HOP and neurodiversity. The guiding coalition members should swap out to provide a continuation of the transformational process with new participants who have a fresh perspective. Additionally, there should be recognition and reward methods put in place to reward employees whose behavior and decisions are in line with the new culture.

At this point the vision has mostly been achieved. The organization should be positioned to maintain the change in a sustainable manner. There should be mechanisms in place to ensure this new way of operating around HOP is embedded into all facets of the organization. This could include policies that are written using

HOP principles as a guiding philosophy. There should also be mechanisms in place to create neuro-inclusive environments and workplace climates so that ND employees do not feel like they are forgotten. If this is done in the right way it will provide benefits to all workers and rather than feeling like something has been "done to them" they will feel like something has been "done with them." In this manner leadership will likely gain more buy-in and support across the workforce rather than only from ND employees. This new change should be intended to increase performance results as well as employee health, wellness and safety.

NOTE

1   Total Worker Health® is a registered trademark of the U.S. Department of Health and Human Services.

BIBLIOGRAPHY

ANSI & ASSP (2019). *ANSI/ASSP Z10.0 -2019 Occupational Health and Safety Management Systems*. ASSP.

Baron-Cohen, S. (2022). *The Pattern Seekers: A New Theory of Human Invention*. Penguin Books.

Cadieux, N., Cadieux, R. (2024) *Operationalizing Neurodiversity for Human Performance*. BCSP Global Learning Summit Webinar.

Centers for Disease Control and Prevention. (n.d.). *About the total worker health® approach*. Centers for Disease Control and Prevention. https://www.cdc.gov/niosh/twh/about/index.html

Centers for Disease Control and Prevention. (2021, May 25). *Total Worker Health in Action: June 2021*. Centers for Disease Control and Prevention. https://archive.cdc.gov/www_cdc_gov/niosh/twh/newsletter/twhnewsv10n2.html

Conklin, T. (2019). *The 5 Principles of human performance: A Contemporary Update of the Building Blocks of Human Performance for the New View of Safety*. Pre-Accident Investigation Media.

Doyle, N. (2020). Neurodiversity at work: A biopsychosocial model and the impact on working adults. *British Medical Bulletin, 135*(1), 108–125.

Dyck, E., & Russell, G. (2020). Challenging psychiatric classification: Healthy autistic diversity and the neurodiversity movement. *Healthy Minds in the Twentieth Century: In and beyond the Asylum*, 167–187. Cham, Switzerland: Palgrave Macmillan.

Fung, L. K. (2021). *Neurodiversity from Phenomenology to Neurobiology and Enhancing Technologies*. American Psychiatric Association Publishing.

Hutson, P., & Hutson, J. (2022). Neurodiversity and Inclusivity in the workplace: Biopsychosocial Interventions for promoting competitive advantage. *Journal of Organizational Psychology*, 23(2).

Kawa, S., & Giordano, J. (2012). A brief historicity of the Diagnostic and Statistical Manual of Mental Disorders: Issues and implications for the future of psychiatric canon and practice. *Philosophy, Ethics, and Humanities in Medicine:PEHM, 7*, 2.

Kotter, J. (2012). *Leading Change*. Boston, MA: Harvard Business Review Press.

Krzeminska, A., Austin, R. D., Bruyère, S. M., & Hedley, D. (2019). The advantages and challenges of neurodiversity employment in organizations. *Journal of Management & Organization*, 25(4), 453–463. doi:10.1017/jmo.2019.58

Landry, L. (2020, December 15). *What Is Human-Centered Design?: HBS Online*. Business Insights Blog. https://online.hbs.edu/blog/post/what-is-human-centered-design

Lazerwitz, Rowe, M. A., Trimarchi, K. J., Garcia, R. D., Chu, R., Steele, M. C., Parekh, S., Wren-Jarvis, J., Bourla, I., Mark, I., Marco, E. J., & Mukherjee, P. (2022). Brief report: Characterization of sensory over-responsivity in a broad neurodevelopmental concern cohort using the Sensory Processing Three Dimensions (SP3D) Assessment. *Journal of Autism and Developmental Disorders*, *54*(8), 3185–3192. https://doi.org/10.1007/s10803-022-05747-0

Mellifont, D. (2020). Taming the raging bully! A case study critically exploring anti-bullying measures to support neurodiverse employees. *South Asian Journal of Business and Management Cases*, *9*(1), 54–67.

Nancy Doyle, & Almuth McDowall. (2023). *Neurodiversity Coaching: A Psychological Approach to Supporting Neurodivergent Talent and Career Potential*. NY: Routledge.

NIOSH. (2020, August 10). *History of Total Worker Health*. Centers for Disease Control and Prevention.www.cdc.gov/niosh/twh/programs/?CDC_AAref_Val=https://www.cdc.gov/niosh/twh/history.html

NIOSH (2024). NIOSH worker well-being questionnaire (WellBQ). By Chari R, Chang C-C, Sauter SL, Petrun Sayers EL, Huang W, Fisher GG. Cincinnati, OH: U.S. Department of Health and Human Services, Centers for Disease Control and Prevention, National Institute for Occupational Safety and Health, DHHS (NIOSH) Publication No. 2021-110, (revised 05/2024), https://doi.org/10.26616/NIOSHPUB2021110revised052024

Pauc, R. (2005). Comorbidity of dyslexia, dyspraxia, attention deficit disorder (ADD), attention deficit hyperactive disorder (ADHD), obsessive compulsive disorder (OCD) and Tourette's syndrome in children: A prospective epidemiological study. *Clinical Chiropractic*, *8*(4), 189–198. https://doi.org/10.1016/j.clch.2005.09.007

Posada, D., & Hebert, J. (2022). *Substance Use Disorders, Cognitive Dysfunction, and Neurodivergence in Emerging Adulthood*. Kennesaw State University, Symposium of Student Scholars.

Pupulidy, I., & Vesel, C. (2023). *Human & Organization Potential*. Dynamic Inquiry LLC.

Ruzich, E., Allison, C., Chakrabarti, B., Smith, P., Musto, H., Ring, H., & Baron-Cohen, S. (2015). Sex and STEM occupation predict autism-spectrum quotient (AQ) scores in half a million people. *PloS One*, 10(10), e0141229.

Singer, J. (1998). Odd people in: The birth of community amongst people on the "autistic spectrum": A personal exploration of a new social movement based on neurological diversity. A thesis presented to the Faculty of Humanities and Social Sciences in partial fulfillment of the requirements for the degree of Bachelor of Arts Social Science (Honors), Faculty of Humanities and Social Science, University of Technology, Sydney.

Whelpley, C. E., Holladay-Sandidge, H. D., Woznyj, H. M., & Banks, G. C. (2023). The biopsychosocial model and neurodiversity: A person-centered approach. *Industrial and Organizational Psychology*, *16*(1), 25–30.

Index